Praise for

JUNIOR SEAU

"Junior Seau shot himself in the heart, maybe to save his head for the pathologist. But Jim Trotter's book is about the heart, and of the heart."
— Tom Callahan, author of *Johnny U* and *The GM*

"This book demonstrates why Trotter is one of the best talents in sports journalism. He didn't just write a great book about a football legend . . . he wrote one of the best sports biographies of all time."
— Mike Freeman, author of *Bloody Sundays* and *Bowden*

"Junior Seau lived a remarkable, passion-filled, and eventually tragic life. His death may make him one of the most important figures in football history. Jim Trotter masterfully tells it all here, in a powerful book that may change the way you look at the game."
— Dan Wetzel, national columnist for Yahoo Sports

"Few people knew Junior Seau like Jim Trotter, who gracefully guides readers into the unique world of one of the NFL's most compelling figures of the last quarter century. Trotter has achieved something rare here: he took a sports book and artistically crafted it into a lyrical narrative about dreams, love, and, ultimately, heart-wrenching loss."
— Lars Anderson, author of *The Storm and the Tide* and *The All Americans*

"Trotter, who covers the NFL for ESPN, tells a difficult story and tells it well . . . This is a powerful, thought-provoking account, handled with grace and sensitivity, of a superior football player's life and death."
— *Booklist*

JUNIOR SEAU

THE LIFE AND DEATH OF
A FOOTBALL ICON

Jim Trotter

Mariner Books
Houghton Mifflin Harcourt
BOSTON   NEW YORK

First Mariner Books edition 2016
Copyright © 2015 by Jim Trotter

For information about permission to reproduce selections from this
book, write to trade.permissions@hmhco.com or to Permissions,
Houghton Mifflin Harcourt Publishing Company, 3 Park Avenue,
19th Floor, New York, New York 10016.

www.hmhco.com

Library of Congress Cataloging-in-Publication Data is available.
ISBN 978-0-544-23617-2 (hardcover)
ISBN 978-0-544-81189-8 (pbk)

Book design by Brian Moore

Printed in the United States of America
DOC 10 9 8 7 6 5 4 3 2 1

To my mother—My biggest cheerleader.

To my family—Thanks for your patience.

To the Seau children—If greatness is making those around you better, your father achieved greatness many times over in his 43 years.

To Junior—Miss you, bud-deee.

Contents

Introduction

Instantly, the room fell silent.

The 1999 NFL season had been over for roughly a month when members of the San Diego Chargers' coaching and personnel staffs met in a second-floor office at their training facility.

They were there to discuss the roster, and one by one, they reviewed each player, identifying his strengths and weaknesses and overall value to the team. Eventually they got to Junior Seau, their perennial Pro Bowl linebacker who had just finished his 10th season.

The conversation didn't last long, but near the end of it, almost in passing, veteran linebackers coach Jim Vechiarella said: "It's not going to end well for this guy."

Everyone was stunned. It was as if a rock had crashed through the window. They looked around at each other as if to say, *Did I just hear what I think I heard?*

Junior was the local boy who made good — a gifted athlete at Oceanside High School and the University of Southern California who went on to be drafted fifth overall by the Chargers in 1990. He was beloved in the community and revered by his teammates. He owned a popular restaurant and had some of San Diego's most prominent power brokers on speed dial.

Not going to end well? Junior Seau?

"It was chilling," said Billy Devaney, the team's director of player personnel at the time. "Vech had been around. He was kind of a crusty old guy. But he wasn't saying it to be negative. It was more out of concern."

"He was worried about Junior and life after football," said former head coach Mike Riley. "Looking back . . ."

Riley paused before continuing: "For a guy like Junior, football is more than an occupation. It's life."

And in this case, death.

On May 2, 2012, only two and a half years after concluding a 20-season NFL career that included 12 Pro Bowl berths, eight All-Pro selections, and two Super Bowl appearances, Junior retreated to a second-floor bedroom in his beachfront home in Oceanside, California, leaned back on a bed, and put a .357-caliber revolver over his heart. Then he pulled the trigger.

The answer to why the 43-year-old killed himself is as complex as the question is simple. There were addictions or dependency to alcohol, prescription meds, gambling, and women. There was brain damage from two decades of violent collisions in the National Football League. There was the suffocating pressure of having to be a human ATM for a community-sized family, not to mention the shame of feeling he had failed because he was headed for bankruptcy despite having left the game with a financial portfolio that didn't require him to work in retirement unless he chose to do so.

Few people knew the severity of his pain or his problems because he never pulled back the curtains to let anyone see it. Instead, he hid everything behind a smile so bright it could light up a stadium. He was a master of manipulation and compartmentalization, often telling people what he thought they wanted to hear — what would make them happy and create the least amount of drama — rather than what he was really thinking or feeling.

He did it with the women in his life, making each one feel as if she was the only one who mattered by telling her that he loved her; in reality, he was saying the same thing to at least one other woman at the same time. He did it with family members, telling them he was unaware that his foundation director and personal assistant had blocked their attempts to reach him or get money from him when, in reality, he was the one telling her not to give it to them. He did it with his trainers and teammates, hiding injuries and seeking physical treatment in private to perpetuate the image of himself as being more myth than man. And he did it with friends, telling them everything was great when, in reality, his life was spinning out of control.

Presenting an image of happiness and control was as important to him as presenting an image of strength. While growing up, he and his five siblings were told regularly by their mother to go into the world and "make happy," which was what he tried to do. His smile was as disarming as his sculpted six-foot-three, 255-pound physique was intimidating. He loved to laugh and sing, even though his voice was often off-key. His playfulness sometimes resulted in comments or jokes that were politically incorrect, but he could get away with it because everyone knew there wasn't a malicious bone on his skeleton. Even Nick Saban and Bill Belichick, coaches whose public personas are as dour as Junior's was jovial, accepted his verbal jabs because they knew the words came from a good place.

The reality, however, is that we don't really know our athletic heroes, as much as we think otherwise. We see only the facade, the parts that they want us to see. This isn't to say Junior wasn't a good man. He was. He gave to those around him without ever asking for anything but a smile in return. Bringing joy to someone else brought pleasure to him. He called everyone Buddy, though in his own unique way. It wasn't BUD-DY, two quick syllables, but *BUD-DEEEE*, with him hanging on to the *DEEEE* as if it were a note in a song.

Countless people would've been there in a heartbeat if they had known he needed help, but he viewed asking for help as a form of weakness. The few who did reach out to him were pushed away or kept at a distance. On those occasions when he felt he had bottomed out, he told people close to him that he was unhappy with himself and wanted to change. But those moments of clarity were fleeting and quickly replaced with more women, more alcohol, and more trips to casinos.

"The two things that he told me that really gave him peace were his children and that surf, to be able to go out there and surf those waves," said Rodney Harrison, a teammate in San Diego and New England. "They gave him that peace for that moment."

His longboard was so big, it could have been mistaken for a kayak, and on most days in the off-season he made a point of walking across the narrow road in front of his home, down the 12 concrete stairs to the shore, then through the sand and into the cold yet comforting Pacific—the same Pacific on which he had body-surfed as a kid.

The weekend after his death some 200 surfers made that same trek and paddled offshore. New Orleans Saints quarterback Drew Brees was among them, despite he and Junior playing together for only two seasons in San Diego. "He was more than just a great football player," Brees said. "He was the heartbeat of the team and the city for a long time. Everywhere you went, he was like a legend. He was like family to everybody. I've never played with someone whom the guys on the team respected more."

That respect extended beyond the locker room. In December 2010, Mark Davis, the freckle-faced, gravel-voiced son of legendary Raiders owner Al Davis, chose to stay overnight in San Diego after a 28–13 defeat of the Chargers before driving to Palm Springs for a brief getaway. He was hungry and wanted to watch that night's NFL game between the Ravens and Steelers, so he went to Seau's The Restaurant.

After passing through the towering glass double doors at the entrance, he was stunned to see the greatest defensive player in Chargers history standing at the front desk. Davis had never formally met Junior, but he definitely was familiar with him. "I've hated you for all my life, but it's out of respect," he said after they exchanged handshakes and small talk. "It's because you kicked our ass so many times."

Junior flashed the broad, welcoming smile for which he was known. Instantly, he and Davis were friends. They retreated upstairs to his private office, where they talked for hours about life and loves, but very little about football.

"I learned that he's such a soft, generous, life-loving person," Davis said. "You talk about alter ego, Clark Kent. I mean, he's in his flip-flops and shorts, just hanging out. It was a great thing. We [met up] again the following year when we played down there. We started a little relationship. He was just arms wide open. I valued his friendship because he was very special. When he did what he did, it was a shock and it really hurt me."

"I was in the car in Atlanta, and I'm listening to the news, and they're reporting Junior's death," said Devaney. "I'm thinking, '*There's no way. Not this guy. This guy doesn't die. He's Junior Seau.*'"

That's what he wanted everyone to believe. Hopefully, this book will help us understand why that wasn't the case—and remind those who are hurting that asking for help is a sign of strength and not weakness.

"I Have to Be Better Than Me"

IT'S EARLY MARCH, and the sun is just beginning to rise over Oceanside, California, a coastal town 45 minutes north of San Diego. Dew is on the grass and a chill is in the air when Sai Niu arrives at the school bus stop at six o'clock. His body is awake only in the sense that his eyes are open.

As he prepares to board the bus, he notices someone running sprints on an adjacent field. He squints through the dim light and walks around the back of the bus to get a closer look. Soon, he realizes it's Junior—or Bug, as he is known to family and close friends.

The classmates exchange handshakes and small talk. They're entering the spring of their junior year at Oceanside High, but already their minds are thinking ahead to the fall and winter, when they will lead the Pirates' varsity football and basketball teams for the second straight year. Junior invites Niu to join him for an early-morning workout later in the week. When the running back/point guard arrives two days later at five o'clock, Junior is waiting for him.

They stretch briefly, then begin running . . . and running . . . and running. Niu thinks he is prepared, but he isn't. His first workout with Junior turns out to be his last. "After that," Niu said of that 1986 morning, "every time the bus would come and I'd see him over there, I made it a point to walk in the opposite direction. Bug would be out there like clockwork. He was one of the hardest-working people I've ever known."

Junior lived in a three-bedroom bungalow where he and three brothers slept in a converted one-car garage that served as his gym as well as his bedroom. On most mornings he'd rise before the sun crept over the coast and exercise until his body was drenched in sweat and his muscles twitched from fatigue. He'd do push-ups and sit-ups on the cement floor, pull-ups on a tree branch in the backyard. He used the neighborhood streets as his personal track.

While running one afternoon, he passed the home of a cousin, who was seated on the front porch. Fifteen minutes later he passed the house again, moving in the opposite direction. The cousin didn't think much of it; Junior was always running. But when the youngster passed the home a third time, the cousin shook his head and laughed. "Man, that kid's crazy," he said to a friend. "But he's going to go somewhere."

The words were prescient. Nearly a decade later, Bug's journey took him to the Super Bowl, where he played on the grandest stage in professional sports. A decade after that, it took him to the White House, where President George W. Bush honored him as a "Volunteer of the Year" for his work with at-risk kids in San Diego County. And in 2015, it took him to Canton, Ohio, where he became the first player of Polynesian descent to be inducted into the Pro Football Hall of Fame.

But to fully appreciate just how far he traveled, literally and figuratively, you must understand where his journey began.

Oceanside is the third-largest city in San Diego County, yet it often is overshadowed by smaller coastal municipalities to the south, like Carlsbad, Encinitas, La Jolla, and Coronado. Those communities are held up as symbols of affluence and privilege when people talk about the beauty of the region. Oceanside is known as the gritty military town on the southern border of Camp Pendleton, the 125,000-acre Marine Corps training facility that's the largest on the West Coast. It's the dirt-covered stone that has yet to be buffed and polished into a priceless gem, three and a half miles of coastline that's as unpretentious as it is gorgeous.

The Seaus did not live close enough to the water to taste the salt in the air. They lived inland, where gangs and drugs and small, overcrowded bungalows were prevalent. Community members referred to it as East Side; while it could be intimidating to outsiders, many locals found comfort there because it was what they knew. Some 1,400 people of Samoan descent resided in the area in 2000, making it one of the largest concentrations of Polynesians in the United States, according to that year's census.

Tiaina Seau and his wife, Luisa, grew up on American Samoa — Luisa in Pago Pago, Tiaina in the much smaller village of Nu'uuli — but they didn't meet until both were in Hawaii, where Luisa was attending school and Tiaina was searching for work. They fell in love, married, and started a family, but thoughts of settling there dissipated quickly after son David was born with a hole in his lung.

The parents were told that David could receive specialized medical care in San Diego, where Tiaina had a sister, so the family packed its belongings and relocated. They spent two years in San Diego before moving 45 minutes north to Oceanside. The change in address stemmed from Tiaina's desire to reduce his commute. He had found work at a rubber factory in San Clemente, and the 90-minute drive in one direction from San Diego was wearing on him. By moving to Oceanside, he could cut the commute in half.

The family settled on Zeiss Street, where children Savaii, Annette, Tiaina, and Antonio joined David and Mary. The baby of the bunch from 1969 to '76 was Tiaina, otherwise known as Junior. Interestingly, he wasn't a true junior. Both his father and his grandfather had the same given name, making him a Tiaina III, but everyone called him Junior to differentiate him from his dad. His mother tended to call him Pepe, which is Samoan for baby.

There's a long-accepted story that the Seau family went back to American Samoa for several years after Junior was born, then returned to Oceanside. It also claims that Junior didn't learn English until he was seven. Neither is true. Junior's first trip to American Samoa didn't come until after his third year of high school. When his family occasionally asked him to set the record straight, he'd shrug and say: "Let 'em run with it. Makes for a better story."

Junior loved to prank people, and rewriting his family history spoke to that. While in college he told a reporter: "I was five years old and couldn't speak English when we came here. But my dad wanted to raise us in America so we could have a chance to go to college."

He was a handful even as a young child, unable to sit still for long stretches and always searching for the next adventure. Mary, the oldest of his two sisters, often had to babysit him while their parents were at work. When his mischievousness would wear her down, she'd allow him to go outside alone, which could be problematic because he tended to stray as far as his feet and his curiosity could take him. No one was overly worried, though, because the 1970s were a more innocent time. Most everyone on the block knew each other, and there was a sense of shared responsibility when it came to watching over the children.

But Junior wasn't the type who needed to be protected from others—he needed to be protected from himself. He'd get into anything and everything. Fear was not in his vocabulary. Once, he and some friends found a mattress in the backyard of an empty house and moved it to the front yard, where they took turns jumping off the roof onto the covered coils. When he did get caught doing something he wasn't supposed to be doing, he had a knack for talking his way out of trouble, something that continued well into adulthood. Oftentimes Mary would find herself in the middle of things.

"When we were young, the boys would build their own go-cart," she said. "They would take the wheels off the grocery carts that people would push into the canyon and use the lumber from the backyard that Dad had purchased to add a kitchen on the back of the house. They knew Dad would be mad when he got home, so they'd ask me to say I was the one who did it. I wouldn't get in trouble because I was a girl, so everything that the boys did they blamed it on me."

Typically, one of their mother's first questions upon arriving home from work was, "Where's Pepe? Where's Pepe?" followed by, "Did he eat? Did he eat?" Junior knew if he told his mom he hadn't eaten—and often he hadn't, because he was busy playing—Mary would get into trouble. So he would lie and then blackmail Mary to get his way later.

The two had a strong bond, not only because she spent so much time babysitting him, but also because she was a good athlete. Mary excelled in basketball and sometimes beat Junior in games of H-O-R-S-E and 1-on-1. This was before he hit a growth spurt, and each time he lost to his sister he'd make some sort of excuse. The two were so competitive that when she was named "Outstanding Female Athlete" at Jefferson

Middle School, Junior didn't congratulate her. Instead, he told her he was going to win the male equivalent when he attended the school, which he did.

When Junior began competing in organized athletics as a sixth grader at the Oceanside Boys & Girls Club, it was apparent that he was superior to others in his age group. Mary was among the first to see it because she spent so much time with him. She had a problem, though: the club basketball team often competed on courts at the beach, where the Seau girls were not allowed to go without adult supervision. She circumvented the rule by telling her folks she was going to communion at St. Mary's, a church that happened to be along the path to the beach. "Each time my friends and I got to St. Mary's, I'd cross my chest and keep walking," she said, laughing.

Junior usually would put on a show. He had the size and strength to go up for rebounds and the speed and crude ball-handling skills to dribble from one end of the court to the other. Even before he was out of middle school his athletic feats were becoming the stuff of legend. "There was one story that he hit a softball so high he was rounding third and on his way home before it started to come down," said Pulu Poumele, a standout athlete who was three years younger.

Because he hated to lose more than he loved to win, Junior developed a reputation for being a bad sport. When defeats occurred, which was infrequently, he tended to handle them poorly. He was known to knock over Connect Four games at the Boys & Girls Club when it was apparent he was going to lose, and after one of his youth basketball teams lost in the championship game of a tournament, spoiling an undefeated season, he ran to the center of the court, sprawled on his stomach, and pounded the floor while crying.

Fights were not uncommon either. In high school he was ejected from a basketball game for slamming an opponent to the ground. Another time, following a loss on the hardwoods, he threw a guy over the counter at McDonald's when he saw the boy flirting with his girlfriend. He later explained his propensity for fisticuffs by saying his older brothers used to take him down the street and have him fight sixth graders when he was in second grade.

"To challenge older boys and win was a big accomplishment," he told the *San Diego Evening Tribune*. Added Bill Christopher, Oceanside's varsity basketball coach through Seau's junior season, "Junior is a fun guy to be around, but he's very intense in games. It's in his blood. Samoan warriors fight to the death."

Mary was present during one fit of anger. Junior was playing in a basketball game as a fourth grader. When things did not go as he wanted, he began throwing a tantrum.

"One of my friends was saying he needed to control his temper," Mary recalled. "At times I'd watch him and be embarrassed, but at the same time I knew he was beating himself. I asked him, 'Why do you get so mad?' He'd say, 'You just don't understand.' I'd tell him, 'I do understand. I play sports.' But he'd say, 'I have to beat myself. I have to be better than me.'"

"He was competing against himself, not the other team," Mary said. "And if his team didn't win, he felt like he hadn't done enough."

Still, for every few stories about him losing his cool, there's another about an act of kindness or compassion. For instance, he'd go door-to-door on his block to get the kids to come out and play, even if they weren't athletic. Sometimes the neighbors declined, but Junior wouldn't take no for an answer. He'd charm and hound them until they changed their minds, then do what he could to ensure they had a good time, so they'd come back.

In one instance, there were two overweight kids who didn't want to come outside. Relentless, Junior finally got them to agree to participate in a relay race. Not only that, he put them on his team even though he knew his side would lose. In situations like that, the outcome wasn't as important as his neighbors' feelings, and he knew they would feel good about themselves because he had demanded that they be on his team.

Finding someone who disliked Junior was difficult. He had charisma and loved to laugh and joke and sing. He favored old-school R&B and would hit the Repeat button so many times that people would get sick of a song long before he did. Slow jams were his favorites.

He seemingly had nicknames for everyone, and when he flashed his broad smile at coeds, they were his. When he put his arm around a teammate's shoulder or patted that person on the back, there was nothing the guy wouldn't do for him. His ability to win over people was neither forced nor contrived. He was one of those unique individuals who had a special knack for relating to people.

The only thing Junior loved more than his music, friends, and family was his high school. Oceanside had a struggling football program when he arrived in 1983, consistently taking a backseat to local powerhouses (and newer schools) El Camino and Vista. That used to gall Seau because some of those schools' best athletes lived within Oceanside's boundaries but circumvented the residency rules to attend the newer schools, including cousin Sal Aunese, a standout quarterback at Vista. Junior had no interest in following in their footsteps, although he could have, because his father worked as a custodian and athletics equipment manager at El Camino. Instead, he wanted to create his own path, damn the challenge.

When he joined the varsity football program as a sophomore, the Pirates had gone eight straight years without a playoff appearance. He missed the first seven games that year because of a broken collarbone, but immediately upon returning he flashed his playmaking ability by returning an interception for a touchdown in the final moments of a 14–7 victory over Carlsbad. The next week, in Oceanside's homecoming against Torrey Pines, he took over after the Pirates' starting quarterback was injured and scored all three touchdowns in an 18–13 win.

Individual success and team success proved to be mutually exclusive, however. The Pirates finished 5-5 his sophomore season and 3-7 the next year. Their struggles went deeper than neighborhood talent choosing to play elsewhere — academics were also a problem. It was common for the Pirates to lose a good swath of the roster when grades came out during the year. That was one reason why Roy Scaffidi, in his first year as coach, in 1986, had many of his trusted core players start on both offense and defense, to lessen the severity of the impact if players were lost to grades. For the plan to work, though, some team members would have to switch positions — including Junior.

Junior heard about the plan while he was out of the country that summer, and he wasn't pleased. He, Niu, and Okland Salavea — three starters on the football and basketball teams — had flown to American Samoa the day after their junior year ended to participate in the Pan

American Junior Championships in basketball. Oceanside was without a football coach at the time, so the players figured it would be a fun way to spend the beginning of the summer. Plus, it was a badge of honor for their parents, who took great pride in their sons' athletic accomplishments.

It wasn't long, though, before Junior began longing for home. Growing up on the East Side of Oceanside was like being in Beverly Hills compared to the spartan conditions of island life. There was almost no water pressure for taking showers, and working stoves and refrigerators were a luxury. The beds were so uncomfortable that the guys sometimes put sheets over their suitcases and slept on the floor. But what really bothered Junior was hearing that Scaffidi was thinking about moving him from quarterback to wide receiver.

Confused and angry, he called his girlfriend, Melissa Waldrop, collect every day, running up an $800 phone bill she paid by working at a local movie theater. He was upset because the change not only might hurt his chance for a scholarship but would also take the ball out of his hands on some plays. Junior loved having the ball in his hands because it meant he was in the middle of the action on every play. He was so disturbed about the possible position change that he cut short his trip by three or four weeks and returned to Oceanside without Niu or Salavea. He didn't tell anyone, but he was seriously considering transferring to Vista High, where his cousin Sal Aunese was on the football team.

Playing quarterback held special significance for Junior because it was the position Aunese played. Junior idolized his cousin. He respected him as an athlete and loved him like a brother. If quarterback was good enough for Aunese, then Junior felt it was good enough for him. But leaving Oceanside was not a legitimate option. The sense of loyalty to his neighborhood school was too strong. It was the school all of his siblings had attended, plus he wanted to finish what he started. He also knew he was good enough athletically to dominate at any position.

After returning from American Samoa, he played wideout in a summer-league passing game and scored four touchdowns. Any thought of leaving Oceanside evaporated at that point. He told Scaffidi he was good with Rocky Aukuso taking over at quarterback, a move that paid immediate dividends. Through the first three games of the season Junior had 16 receptions for 242 yards and two scores. He also was excelling on defense, where he had been moved from safety to outside linebacker.

"We think that's where his future is, where colleges would want him to play," Scaffidi correctly told the *North County Blade-Citizen* early in the season. "I expect him to be an All-America in college. If he's not an All-America in high school, then I've been coaching in a cave somewhere. I've seen some great football players and they don't come any better than him. He's the type of player that makes you a great coach. He's one of those incredible athletes that happens once every 10 years."

"I don't think we knew how good he was until we got out of school," said Niu. "He was a monster on the field, hated to lose, and would fight anyone. The way he acted, we had to pick up our level to play with him too. We'd do hitting drills, and it didn't matter if you were a little guy. He'd try to rip your head off. He made us better like that."

Junior just wanted to win. Losing ate at him for days at a time. "I dreaded when they lost. It was almost like I didn't want to be around him," said Melissa Waldrop, who began dating Junior when he was a junior and she was a sophomore. "He would take it so personal. He couldn't just brush it off his shoulders. It would affect his mood. Not just for that night, but sometimes for days or until he got back out on the court or field to redeem himself."

Scaffidi's bold moves resulted in the Pirates' first winning season in nine years and first playoff appearance in 11. And though they were routed 41–7 in the section final by a loaded squad from Lincoln Prep, former home to NFL greats Marcus Allen and Terrell Davis, the significance of the season could be appreciated only over time. It proved to be the launching point for 28 consecutive playoff trips—and counting.

Junior ensured that success by advocating for local kids to stay home and play for the Pirates. When Poumele was an eighth grader, there was speculation about which high school he would attend. He was a talented athlete who lived within Oceanside's boundaries, but his father was a minister at a church in Vista. Getting around the residency rules would not have been difficult.

Junior did everything he could to persuade Poumele to wear the green and white of the Pirates. "He said, 'Vista and El Camino, they've got everybody set. You've got to make sure you come to Oceanside. We're going to build something here. It's going to happen,'" recalled Poumele, who attended Oceanside before playing collegiately at Arizona. "I probably would've gone to Oceanside anyway, but his words meant a lot."

Junior was so loyal to the Pirates that he skipped the California State Track and Field Meet as a senior — he was favored to medal — because it conflicted with a school luau to raise money for the athletics department. Junior had committed to doing a native Samoan dance at the event and didn't want to go back on his word. It was another example of the increased maturity people noticed in him that year. Part of it could be traced to an incident on the basketball court the previous winter, when he body-slammed an El Camino player. Junior's father, who always preached the importance of honoring the family name, was embarrassed and angry. As a deacon in the family church, he felt that Junior's actions undercut his credibility among parishioners. So he made Junior apologize to the Wildcats' players and coaches.

"I learned a lot from that El Camino game," Junior told the *Blade-Citizen*. "It's better to learn it now than to learn it later. They're not going to cope with it [at college]. They'll just send you home. Knowing that this is my last year, I want to leave [on] a good note and at the same time show examples for the younger kids under me. They look up and I feel that I'm a role model. They elected me captain of the team, and I've got to live up to it. I didn't ask for it; they picked it for me."

"I was told that one of the things that I need to be concerned about was that he sometimes plays out of control," Scaffidi said of Junior. "Well, it may be just a maturing process, but — and I would not take credit for it — I think somehow between last year and this year he decided to keep his emotions under control. And so far he's done everything I could possibly expect."

Junior was starting to see the big picture, to understand that sports were not just about fun and games. For someone like him, sports could be the key that opened the door to a prosperous future.

From the time Junior first exhibited superior potential as an athlete, exceptions were made for him that were not made for his siblings. His brothers and sister had to get summer jobs to help with family finances — it's a tradition in Polynesian families for teens to contribute toward the household finances — but Junior did not. His siblings had multiple household chores, but Junior did not. He was permitted to focus solely on school and athletics, with the hope and expectation that he would be the one to lift the family out of poverty if he made it. There was no resentment on the siblings' part. They were proud of his success and they knew what was at stake.

"My wife said, 'He should go look for a job,'" said Papa Seau, "and I told her, 'Let him go play.' Sometimes people asked, 'Why doesn't Junior work, contribute to the family?' I said, 'That's okay.' His mind, his thought, was sports, and I didn't want to stop it."

Money was never abundant in their home. While his father worked as a custodian and equipment manager at El Camino High School, his mother worked in the commissary at Camp Pendleton and at a laundromat. Consciously or not, Papa Seau pushed his son hard. He greeted victories with smiles and a few dollars for spending change, but losses were met with a sort of dismissive silence. The tough love hurt the sensitive youngster, who, like any kid, wanted validation from his father.

"You know what they say about silence: it's deafening to the ears and to the soul," said Waldrop, who would remain Junior's girlfriend through college. "When they lost, he didn't want to go home. He didn't want to face the silence or disappointment. I was never there when he went home to experience that; I only know what he told me, and that was that the silence was more painful than if his dad was to yell at him. I don't know if that was self-perceived. His dad is a very quiet man and reserved, to the point that when he does speak, it's going to be heard. People are going to listen. So for him to be quiet wasn't out of the ordinary, but I know Junior felt pain from it. Whether it was put upon him or his own preconceived feelings, June felt that he had to be the best. When they would fail to win a game, he'd feel like he hadn't done enough to help his teammates. To him it was never a team sport unless they were celebrating."

Because his job as equipment manager required him to attend the Wildcats' games, Papa Seau typically could see his son play only once a year in football and twice a year in basketball — when Oceanside faced El Camino. Mama Seau was at every game. Even when Junior told her not to drive long distances to away games, she was there. He could hear her deep voice or her piercing blow horn through the din. The combination of her voice and his father's silence helped push him to become the greatest prep athlete in San Diego County history.

As a senior in football, he had 62 receptions (which tied for fourth-most in county history) for 891 yards and 11 touchdowns and was the "San Diego Section Defensive Player of the Year" at linebacker. *Parade* magazine named him an All-America at "athlete," a designation it created especially for him. In basketball he averaged 22.3 points and nine rebounds a game and was named "Section Player of the Year." In track and field he won the Avocado League championship with a shot put of 53 feet, 5.5 inches, and broke an uncle's school record with a mark of 56 feet, 10 inches. His true love was basketball, but he knew there weren't many six-foot-three-inch power forwards in the NBA, so football was where he planned to make his mark.

"I'm going to be a professional football player one day. Just you wait. You'll see," Junior told Waldrop as an eleventh grader, stopping between the auto shop and the gymnasium to look her in the eyes and make his point. She smiled and was supportive, but in the back of her mind she was thinking: *What talented young athlete doesn't think that?* But she also saw a purpose in his eyes, something that said this wasn't wishful thinking.

"He knew what he aspired to be," she said. "He wanted to take care of his parents and family."

Junior was keenly aware that they had made sacrifices for him — not just his parents but also his siblings. He felt he owed his sister Mary, who paid the registration fee the first time he played football and often provided him with money to purchase cleats and other items. She also helped him purchase the promise ring that he gave to Waldrop. His parents had spent money on him that took away from contributions they might have made to their community church and extended family. Sharing responsibility by pooling resources is highly valued in Polynesian culture, but while it is a beautiful idea in theory, it can become burdensome in practice. Junior would soon learn this.

Success and Shame: One and the Same

JUNIOR ALWAYS LOVED being the Big Man on Campus in high school. At six feet, three inches, and 215 pounds, his shoulders were broad enough to carry the responsibility that came with it. Or so he thought.

Midway through his senior year, his knees began to buckle from the weight of being one of the most sought-after high school football players in the country. College head coaches were showing up at his basketball practices and games, and recruiters regularly waited in cars outside his home or called the campus pay phones in search of him. Ohio State coach John Cooper was preparing to lead his Buckeyes onto the field for the Rose Bowl when he stopped and phoned Junior from the locker room. Things were so crazy that Junior spent some nights at an uncle's place to find peace. Other times he slept at his girlfriend's house.

Don Montamble, Oceanside's first-year basketball coach, realized something was wrong during a midseason practice. Junior was supposed to flash to the high post, catch the ball, then turn and make an entry pass to the low post, where a screen had been set for the center. Instead, Junior took the pass, turned, and launched a 20-foot fadeaway jumper. Montamble blew his whistle. "Do it right," he said. "Run it again."

Junior shot an irritated look at him that was completely out of character. He always had been respectful of his coaches, regardless of how he felt about them. But this was clearly an attempt to show up Montamble. The coach let it slide until Junior flashed to the high post, caught the ball, and shot a *What now?!* look in Montamble's direction after putting up another 20-foot fadeaway jumper.

"I was incredulous," Montamble said. "I blew my whistle again and was yelling as I walked toward him."

Junior didn't retreat. To the contrary, he began walking toward Montamble, who was only eight years older than him. Everyone in the gym froze in astonishment. *Is this really happening?* they thought to themselves.

Separated by inches, Montamble screamed in his star player's face. Junior said nothing, but his anger was obvious. His nostrils tended to flare when he got mad, and by this time you could fit a basketball up them. Junior tore off his reversible jersey and threw it down. With Montamble still railing, Junior turned and walked toward the exit.

"If you take one more step through that door, you're done!" Montamble shouted. "You won't play another minute of Pirate basketball the rest of your career here, so you better think about that!" He then turned to the team, put a reserve in for Junior, and told the players to run the play correctly. He was watching his players, but his mind was racing with the potential fallout of a 26-year-old first-year coach booting the star player off the team. But he also wanted to remain strong. There had to be boundaries. *You know what's right,* he said to himself. *No athlete can pull this crap without accepting the consequences.*

Junior stood in the doorway, facing the exit, for at least five minutes—until the team finished its drill and started shooting free throws, which typically signaled the end of practice. Then he turned and walked to the bleachers, where he sat by himself. Montamble knew he had Junior the minute he returned, but he wanted to push a little harder because he viewed it as a teachable moment. He walked over and sat next to him without looking at him. "The next thing that happens is, you don't get another second of play time until you apologize to this team," he said. "We're focused on winning a championship, and you pull this shit? It's inexcusable, and you're not going to play until you apologize."

Montamble got up and walked away. If he knew Junior as well as he thought he did, he figured the multi-sport star would cool off and address the team at some point—if not that moment, then later. It was the day before a game, which meant mandatory postpractice video study. When everyone arrived in the room just off the gym, Junior was sitting in the front, waiting. He typically was the last player to arrive, so this got everyone's attention. Montamble began to speak to the team when Junior interrupted and asked to address the squad. When the coaches started to leave, he asked them to stay. Suddenly, tears were rolling down his cheeks. "I sincerely apologize to all of you for what I did," he said. "It's something I would never encourage anybody to do. It was wrong. It's something I know I shouldn't have done. I need to be punished, and I'm willing to accept anything that Coach wants to levy in terms of punishment. Don't think bad of Coach; I put him in that situation, and it's not going to happen again."

Montamble broke the meeting and everyone left — except Junior. For the next five minutes, coach and player cried behind the closed door. Junior revealed that the pressures of being recruited were too much for him. He thought the attention and courtships would be fun, but now he wanted to hide. He felt like a grenade whose pin had been pulled, and it was a matter of time before his emotions exploded. Montamble's heart raced with empathy and anger. He thought: *Someone needs to help this kid! Why has no one helped this kid?!*

Junior's parents had never been through the process and were unaware of the pressure their son was feeling. Montamble figured that since he was spending more after-school time with Junior than anyone, he would take a mentorship role. He and Junior sat in the room and mapped out a plan. First, Junior would use the weekend to narrow his list of five finalists: Colorado, USC, UCLA, Texas, Arizona State. Second, there would be new rules for recruiters: no contact during the day until practice was over, and no contact at all on game days.

UCLA and Texas were the first to be erased from the list, and Junior immediately felt a sense of relief. Texas was too far away, so eliminating the Longhorns was easy. UCLA was trickier. Almost from the beginning his father had been supportive of the Bruins. They had won five straight postseason games under Terry Donahue, including three Rose Bowls. They also were close to home. But Junior didn't hit it off with Donahue. He never said what it was, but something didn't feel right. So he took UCLA off the list without notifying the school. When word trickled back to the Westwood campus, a Bruins recruiter showed up on the Oceanside campus. He was extremely angry, if not belligerent. He told Montamble that he wanted to see Junior.

Basketball practice was about to start, and Montamble reminded the recruiter of the new rules. Didn't matter—he wanted to see Junior immediately. "Not going to happen," Montamble said. Words were exchanged, some not so nice. Montamble turned, walked into the gym, and shut the door. If UCLA needed confirmation that it was off the list, it now had it.

That left Colorado, Arizona State, and USC—until two Arizona State coaches showed up at Junior's basketball game at San Pasqual. Whether they knew about the new guidelines prohibiting such contact is unknown. But they thought they were okay because Scaffidi, the football coach, brought them to the game. Junior spotted them and immediately felt anxious and irritated. The weight that had lifted from his shoulders was back.

"I'm in the gym being interviewed by the media, and my assistant's in the locker room supervising the kids, like he always did, and Junior comes out of the locker room into an empty gym and starts knocking over chairs and screaming," Montamble said. "I had to calm him down. Normally the players all ride the bus back to school, but that night I had him go out a back door so his family could drive him home." Arizona State was definitely off the list now, even though Junior didn't say so.

That left Colorado and USC. Colorado was the clear front-runner in Junior's mind. The Buffaloes had done a great job tilling Oceanside's fertile soil for Polynesian talent, and Junior loved the idea of reconnecting with his cousin, Sal Aunese, and former Oceanside teammate Okland Salavea, both of whom were already at Colorado, a year ahead of him. In addition to being a star quarterback at Vista High, Aunese had competed with Junior on their church team and at the Boys & Girls Club. Salavea had been a standout linebacker at Oceanside and played center on the basketball team. His girlfriend also was close to Waldrop, Junior's girlfriend.

When Junior returned from a recruiting visit to Boulder, he told some friends and family he was going to attend Colorado. It might have been the worst-kept secret in recruiting, and yet, when National Letter of Intent Day rolled around, he did not sign with the Buffaloes — or with anyone else.

The long-standing story is that his father stepped in and quashed the idea of him going to Boulder because it was too far away and the family wouldn't be able to see him play. Not true. Papa Seau actually told Junior that he'd support whatever decision he made, though he stressed that he wanted to be able to see him play (UCLA was his first choice). Mama Seau was the one who cautioned Junior that he'd be far from home without any immediate family. She didn't tell him he couldn't go; she just told him to seriously think about what it would mean if he did.

When news reached the USC offices that Junior was still on the market, the coaches were ecstatic. "I'll admit, I was nervous as hell going into signing day," said Gary Bernardi, who was recruiting Junior for the Trojans. The staff worried that his signing with the Buffaloes . . . or someone else . . . was a fait accompli. Junior had not made an official visit to the USC campus, even though several weeks earlier he had promised new coach Larry Smith and Bernardi that he would do so. Deep down, Bernardi kept holding on to the fact that Junior had given him his word he would visit the university. He believed that Junior was a young man of integrity, but he also knew the realities of the recruiting world. USC had lost track of Junior for a two- or three-week period as it transitioned from the fired Ted Tollner to Smith, who was coming over from the University of Arizona with barely a month to go until signing day. Bernardi, who had been recruiting Junior to attend Arizona, suddenly had to switch gears and sell him on the virtues of USC, which had managed only one bowl victory over the previous seven years. The USC staff had every right to believe it was out of the picture.

But Junior proved to be a man of his word: the day after not signing with anyone, he agreed to make a visit to USC that weekend. "At that point the coaches pretty much got everybody involved," said future Chicago Bears safety Mark Carrier, one of Junior's hosts that weekend. "We put a full-court press on him. It was like, 'If we get this guy on campus, we can't let him leave.' When I met him, I remember thinking, *Holy smokes. Ain't nobody on our team who looks like that.*"

Junior needed to make a decision. The process had dragged on long enough. So he agreed to one final home visit from each of the schools on Tuesday and Wednesday. Whether by coincidence or strategy, the Trojans were the last to see him. Junior called his father and told him not to leave the house because Smith and Bernardi were coming by. "Smith told me and Junior, 'I don't care what school you go to, I still support you guys and I still love you guys,'" Papa Seau recalled. When the meeting was over, the youngster asked his parents to wait inside while he walked Smith to his car. The gesture got their attention because Junior had not walked any other coach beyond the front door. When he returned, he told his mom and dad he was going to sign with the Trojans. He didn't tell that to Smith at the car, but he did break the news to him the next day.

The celebration proved to be short-lived, though, as Junior later failed to achieve the required qualifying score on his college entrance exam to be eligible to compete as a freshman. He fell victim to Proposition 48—NCAA legislation implemented in 1986 that required incoming student-athletes to have at least a 700 (out of a possible 1,600) on the Scholastic Aptitude Test (SAT), or a comparable score on the American College Test (ACT), as well as at least a 2.0 (out of 4.0) grade point average in 11 core high school courses. The legislation was an attempt to improve graduation rates among student-athletes, which among Division I scholarship players in the early 1980s was 33 percent for basketball players and 37.5 percent for football players. *Sports Illustrated* reported that the graduation rates for black student-athletes in those sports were even lower: 29 percent for the freshman classes of 1984 and '85, the last classes to be admitted before Prop 48 was enacted.

Critics panned the legislation because they believed it would negatively impact minorities from disadvantaged backgrounds. They claimed the entrance exams were culturally biased at best, and racist at worst. Junior had the grades, with a 3.6 average, but not the test score. He took the SAT multiple times—even once verbally because Samoan, not English, was considered the first language in his home—but still came up 10 points shy of the required score.

He was at school when he learned that he had come up short on his final attempt. He called his sister Annette and asked if their parents were at the house. When she said yes, he told her he was going to call right back and asked her to stay by the phone. He wanted her to be there when he broke the bad news to his father.

"Education was the main thing in this house," said Annette. "First there was God, then family, then education. Sports came later. When Junior told Dad he didn't pass the test, Dad was pretty upset. Junior had to calm him down. When he did, Dad said, 'Do your school and get ready for next year.'"

Junior felt like a failure, and it weighed on him. He thought he had shamed not only his family name but also his school and his community. People talked as if his 3.6 GPA had been handed to him. He even publicly apologized to the student body during an assembly, and later he went into semi-seclusion.

"He didn't want to show his face," Waldrop said. "The shame, the responsibility that he felt—it broke my heart to watch him go through that. He would hide at my house."

If the silence he heard from his father after losses was unnerving, the treatment he received after failing to earn those final 10 points was tragic. The Polynesian community in Oceanside is extremely close. The church serves as ground zero for many events, and First Congregational Christian Church of Oceanside (though located in Vista) is where many husbands and fathers gather to discuss the topics of the day, including their children. For Junior's father, a proud man whose grandfather was a village chief in Pago Pago, it was difficult to accept that his son had not fulfilled his responsibilities and that friends and outsiders were looking down at him.

"When my dad realized that people were talking, things got worse," Annette said. "They were saying things like, 'Whatever happened to his son? His son was supposed to be the best.' It got really bad. He had to go to the church and hear the talk in the back of his ear, 'Oh, how proud of his son is he now?' It was hard for both of them, but especially him because he's a deacon and he preaches. He had to keep a straight face and do the work of God and support Junior. But Junior took it really hard."

"Nobody stuck up for me — not our relatives, best friends, or neighbors," Junior later told *Sports Illustrated*. "There's a lot of jealousy among Samoans, not wanting others to get ahead in life, and my parents got an earful at church: 'We told you he was never going to make it.'"

When he left for USC, the only thing he received from his parents was a hug from his mother. Melissa hosted a barbecue for him and some close friends, but that was it. The two of them sat on the tailgate of his truck, his personal belongings in the back, and said their good-byes through tear-filled eyes.

Once on campus, he felt isolated and alone because Prop 48 guidelines prohibited him from practicing, attending meetings, training, or eating with the team. He'd call Melissa every day, and she'd sit on the floor in their dining room — her family didn't have a cordless phone — and talk with him for hours. "When are you coming? When are you going to get here?" he'd ask.

Melissa was a year behind Junior but planned to graduate a semester early and move to Los Angeles, where they would get a place together. Until then, they had to settle for him returning on the weekends. Sometimes he'd stay at her house without telling his parents he was in town. The pressure and shame he felt was that significant. So was his fear of feeling insignificant. On multiple occasions, before visiting the house, he'd call to see if his dad was home. If he was there, Junior would ask Annette to meet him someplace else so they could visit. Or he'd wait until his father was gone so he could see his mother.

"My dad was very embarrassed," said Mary. "He didn't even talk to Junior. It killed Junior not being addressed. It killed him that other people around him, people that he trusted, didn't stay in touch with him and believe in him until he started playing. He told me on the phone one day, 'Forget these people who don't believe in me.'"

There was little solace on campus either. Junior's contact with the players was minimal. "You didn't really think about him because he wasn't around," said outside linebacker Michael Williams, who in 1989 started at outside linebacker and roomed with Junior on the road. "You knew you had this guy who was sitting out and would be a part of the team at some point, but when you're practicing day in and day out, you're focused on the guys around you."

Typically, Junior left the weight room just as the team entered it. Sometimes a player would mock him—"Look at that dumb jock"—but he never reacted. Not until his patience wore thin one afternoon when someone included a derisive remark about his heritage in the comment. "That's it," Junior calmly replied. "Meet me outside."

Junior ended the fight fairly quickly, though he broke his hand in the process. Teammates contended that they weren't sure which player was involved in the fight, but the bread crumbs appeared to lead to linebacker Craig Hartsuyker, who admitted to getting into a scrap with Junior.

"We were in the weight room or something," Hartsuyker said decades later. "I couldn't tell you what it was about—certainly something stupid. Back then you're 20 years old and everything seems important, I suppose. Knowing my personality and knowing Junior's—and this is not to lay blame anywhere—but he always was an emotional guy. I might have said or did something that set him off."

In between working out, attending classes, studying, and knocking people out, Junior longed for Waldrop. She graduated from Oceanside High a semester early, in January of 1988, and loaded up her VW Bug and moved to LA, where she rented a studio with the help of her parents, who also gave her some old furniture. She was able to transfer from her job as a sales associate at a clothing-store chain in Oceanside to a store in Westwood, and later she enrolled at a fashion institute in downtown LA.

Junior wasted no time moving in with her, unbeknownst to his family. He needed Waldrop's support, that sense of feeling like he mattered to someone, but he kept the news from his parents because he figured they would disapprove and he didn't want more drama. "God love her," Carrier said of Waldrop. "She pretty much took care of him. She was his chauffeur and just helped him get acclimated to being away from home."

By the following summer, Junior was like a bear emerging from hibernation. He began to show his face more consistently at home. Football was approaching, and he was eager to get on the field. He had already been a physical specimen when he left for USC, but a year in the weight room had morphed him into something frightening. He had gone from 215 pounds to 245, with much of the new muscle in his neck and shoulders. Said Poumele: "When we saw him, we were like, 'What the hell have you been doing up there?'"

The football world was about to find out.

3

"There Was Nothing Junior About His Game"

From the moment he got on the field you knew you had somebody special. He was edgy. It was a difficult adjustment for anybody in college to match up against somebody who already had a professional's body. And he only knew one speed: fast.

— FULLBACK LEROY HOLT

WHEN JUNIOR FINALLY stepped onto the football field for his first practice with the Trojans, he was like a caged animal that had been set free. He had never gone a full year without participating in an organized sport, and it made him realize just how much he missed football. At the same time it also made him hungry to show people what *they* had missed while he was out: a truly gifted linebacker.

Success would not come immediately, however. Unlike in high school, he wasn't joining a losing program that had suspect talent and limited participants. He was joining a program that returned four of its top seven linebackers, in terms of tackles, from a squad that had won the Pac-10 title and advanced to the Rose Bowl the previous season. Further complicating matters for him was that each returnee was familiar with the scheme while he wasn't.

After missing spring ball for two years — in 1987 because he was still in high school and in 1988 because Prop 48 rules prohibited him from practicing with the team — Junior's athleticism, energy, and physicality jumped out at onlookers when he got on the field for fall practice in 1988. Unfortunately, so did his mistakes. He wasn't sound from a schematic standpoint, and a badly sprained ankle forced him to miss the first three games and slowed his development.

He was so raw that the coaching staff limited him to special teams and situational pass-rush duties. It was the first time in his athletic life — with the exception of a brief spell in high school, as a sophomore on the varsity basketball team — that he was a backup. Worse, he finished sixth among linebackers that year, with just 35 tackles.

"He was so physically gifted that he'd destroy guys in one-on-one situations, but we ran a regimented, gap-control-style defense, and he didn't really fit those concepts," said defensive end Tim Ryan. "They kind of had him out of position."

The lackluster showing ate at him. He was accustomed to being the best player on the field, not just the best linebacker on the field. Hartsuyker wasn't nearly as physically gifted, but he still finished with twice as many sacks (eight) and forced fumbles (four) as Junior did. Things were so bad late in the year that Junior was benched after five plays in the UCLA game. It was embarrassing and humbling for him. *I have to beat myself. I have to beat myself.*

To the coaches' credit, they were able to look past the mistakes and recognize that they had a special talent who was capable of making game-altering plays. But how best to get peak performance out of him? They figured it out in the spring of '89, when they moved him to the weak (open) side of the formation, where there would be fewer roadblocks to chasing down the quarterback. Prior to that, he had been playing on the strong side, where he usually had a blocker directly in front of him and was required to read an offensive lineman's movement before pursuing the ball. As a result, he played slower than normal because he either got caught in traffic or was thinking too much.

The staff wanted to find a way to turn him loose, and secondary coach Bobby April saw a means to that end while watching tape at the Los Angeles Rams' practice facility after the season. The Rams were coached by John Robinson, one of the great field generals in USC history. He had two stints at the school, the first from 1976 to 1982 and the latter from 1993 to 1997. During his initial tenure, the Trojans finished second in the final Associated Press poll three times in his first four seasons and earned a share of the national title in 1978.

While with the Rams, Robinson had an open-door policy for USC coaches. They could visit anytime they wished, which was all April needed to hear. He was a young assistant eager to learn, and he camped out in the Rams' video library. One of the things he noticed while studying tape of the Minnesota Vikings' 36–24 victory over the San Francisco 49ers in a divisional playoff game on January 9, 1988, was that USC essentially was running the same defense as the Vikings, with Junior playing the same position as Minnesota's Pro Bowl defensive end Chris Doleman. There was one major difference, however: Doleman didn't have to engage and read blockers at the line of scrimmage before pursuing the ball.

"He was lined up on an angle to that [open] side and flying up the field two yards behind the tackle and going to the football," said April, who also noticed that former USC safety Joey Browner was a member of the Minnesota defense. April called Browner and asked if they could meet so he could pick Browner's brain about how the Vikings were utilizing Doleman. Browner agreed, and they spent an afternoon going over various looks and responsibilities until April and defensive line coach Kevin Wolthausen felt comfortable with what they had learned. The two took the information back to USC, and a new day had dawned for Junior.

"When they tweaked the defense for Junior, he took off," Ryan said. "They basically turned him loose and said, 'Get after the quarterback.'"

At his new spot Junior had more freedom and fewer blockers to deal with. If there wasn't an offensive lineman in front of him in passing situations, for instance, he could drop into a three-point stance and play rush end. His sole job: get up the field and find the ball. His quick first step gave him an advantage over many of the heavy-footed tackles trying to block him because he often got past them before they knew what was happening. His presence forced the University of Washington to go primarily with three-step drops that year because the Huskies feared a sack or turnover if Cary Conklin held the ball longer than that.

"Junior was changing the way teams had to protect because they had no answer for him coming off the edge," Carrier said. "For us, it was easy to adjust and fit him in because we were a veteran group. We had a lot of three- and four-year starters, and we'd play off of him. I had seven interceptions that year largely because I played on his side. I was no dummy. I knew the ball was going to be coming out fast, so I'd go get it."

Junior took to the new defense so quickly that the coaches moved him into the starting lineup during spring practice, ahead of Hartsuyker. Mind you, Hartsuyker was coming off a year in which *Football News* had selected him to its sophomore All-America team. He was no slouch. However, he played the final six weeks of the '88 season with a stress fracture in his left foot, and after a 22–14 loss to Michigan in the Rose Bowl, he wore a cast for three months and missed spring practices. That opened the door for Junior, who was there every day, taking advantage of the opportunity.

But just when he seemed to be prepared to take off, he suffered a flat tire. On the first day of fall practice he sustained a compound fracture of one of his fingers. He couldn't help but think back to his sophomore season of high school, when he sprained an ankle on the first day of fall practice and missed three games. Here he finally had earned a starting job, on a defense reconfigured to take advantage of his skill set, and now he might be out again?

There was no way he was going to let that happen. He had the trainers wrap it up, and he took medication for the pain before and after practice. Then he went out on the field and performed. Performed well.

In a season-opening loss at Illinois, Junior had seven tackles. In a 19–0 win over Stanford, he played four different positions and had three sacks. In a 48–6 demolition of Oregon State, he had three and a half sacks. In a 42–3 spanking of Ohio State, he had two sacks. Still, his best outing arguably was a 24–3 victory over Arizona, in which he had one sack and five unassisted tackles for minus-23 yards. "He was an animal," wideout John Jackson told the *Los Angeles Times*. "They couldn't do anything. It was comedy because they couldn't run a play. He one-man-showed it."

The coaching staff was concerned entering that game. Arizona was averaging 32 points over its previous four games while utilizing a wishbone offense. The run-based scheme used a lot of misdirection and sleight of hand, which put great pressure on defenders to be disciplined and sound in carrying out their responsibilities. There were questions about how Junior would handle it, because the Wildcats figured to use his aggressiveness against him — to turn his strength into a weakness.

"He was everywhere," April said. "He disrupted everything. They couldn't do anything. They had an offense that gave you a lot of problems, and Junior solved all of them. He just wrecked it."

Tom Roggeman was the Trojans' linebackers coach at the time. He's now in his 80s and likes to say that his memory isn't as clear as it used to be. One thing he's certain of, however, is that building their defense around Junior was the right move. "If you have a running back who can get you yards and touchdowns, you play him. It's no different on defense. We had this gem named Junior, but there was nothing junior about his game. When it came down to being a fierce competitor, he had it in bundles."

Sometimes too much of it. Just like in high school, he found himself in multiple scraps on the field. He was even ejected from a game against Cal when he and offensive lineman Steve Gordon got into a scrap. "It was just one of those stupid things," he told reporters. "I've got to learn from it. I could have backed out. I should have. I saw two of their guys on one of ours, and I just went over to make sure he was OK. I got into a little [tussle]. A lot of guys were involved, but it seemed like I stuck out. I felt the officials used me as an example to settle things down."

Coach Smith also used him as an example to the team. Smith was livid in the postgame news conference, despite his team assuming sole possession of first place with the 31–15 victory. He didn't like the 12 penalties for 137 yards and barked about it to the media. He also told Junior that if he got into another fight he would kick him off the team or suspend him. An idle threat, perhaps, but Junior took it so seriously that when a scrum broke out the next week against Notre Dame, he was nowhere near the center of it. It killed him too. "He was so loyal and dedicated, like he was with family, that if he saw one of his teammates in trouble, he was going to go defend them," Carrier said. "But against Notre Dame all he did was pull bodies off each other. He wasn't getting involved in anything."

There was too much at stake. Junior was now thinking about bypassing his senior season and turning pro. The NFL was just beginning to allow underclassmen to enter the draft without first showing some sort of hardship, and Junior wanted to cash in. He and Williams, his roommate on the road and fellow linebacker, would sit on their beds some days and watch NFL highlights on TV. On occasion Junior would point out guys he had beaten the previous year in college, or guys whom he thought he was better than at that point. If those guys could make it in the NFL, why couldn't he?

He finished the year with an astounding 19 sacks and 27 tackles for loss. He was named first-team All-America and the Pac-10 "Defensive Player of the Year." Ryan finished with one more sack, but "mine were of the pull-down-from-behind variety," Ryan said. "His were 'wow' sacks. After that first game he just started smoking people. It was just speed and power. It wasn't real technical."

The weekly grades confirmed as much. According to teammates, Junior often graded out lower after games than every other linebacker when it came to fulfilling his assignments. However, he also was regularly at the top of the list when it came to explosive plays. "He would throw the concept of the defense out the window if he saw something he could take advantage of," said Williams. "The coaches would be like, 'What the fuck are you doing, Junior?! That's not your responsibility!' The next play he would make up for it and they'd be like, 'Great play, Junior! Great play!'"

Junior was so good that, after he turned pro, the USC coaches started assigning the "55" jersey to incoming linebackers whom they believed could be great leaders and dominant players for them. It came to be known as the "55 Club" and was an incredible gesture considering that Junior had started only one season and played just two overall. USC also wasn't some gimmicky program. It ranked among college football's blue bloods, producing future Pro Football Hall of Famers like running back Marcus Allen; defensive back Ronnie Lott; offensive tackles Bruce Matthews, Anthony Munoz, and Ron Mix; and halfback/

flanker Frank Gifford. None of them had had his jersey number ordained in such a fashion.

"I was ecstatic," said Ryan. "I thought it was something that should've been done, something that *needed* to be done."

Junior also had another reason to turn pro: he was a father. Melissa Waldrop had given birth to Tyler Christian Seau in June of that year, before the season.

"He was proud when Tyler was born," said Waldrop, who moved back in with her mother early in the pregnancy and remained there after the delivery. "He was excited to have a son. He said all along he was going to have a son. He picked out his name. Why Tyler? It's what he wanted. He never gave a specific reason. I picked Christian.

"I tried to make it really easy for him during the year [Tyler was born]. Junior's place for our future, for our family, was to stay in school. There was never the option to leave school and get a job to be a provider. That was my job at that time. His was to do those things later."

Later suddenly meant now. But instead of life getting easier, it became more complicated. A new woman was about to enter his life.

4

Ferdinand the Linebacker

ONE OF THE hallmarks of a good agent is his ability to close deals, and Steve Feldman has closed more than he can remember over his career. But this one was different because it had nothing to do with incentives or bonuses or annual salaries.

"There's this great kid I have who's coming out for the draft, and I think you should meet him," Feldman said to Gina DeBoer, who was working in the Chargers' executive offices as an account executive.

Feldman and DeBoer had developed a professional friendship from his frequent visits to the Chargers' offices. He was the gregarious, older, playful type, while she was a slender, attractive blonde with big hair, freckles, and a welcoming smile.

"I think you'd really like him," Feldman said. The kid to whom he was referring was Junior.

DeBoer chuckled and shrugged it off as playful Steve being playful Steve. Later, she walked out of the office and saw a muscular young man in shorts, a tank top, and flip-flops being interviewed by the media. It was Junior. He stopped midsentence and called out "Hi" to her.

Slightly embarrassed, she smiled and kept moving. When she returned, Junior was still being interviewed; again, he stopped midsentence and called out "Hi" to her. She smiled again and kept walking, but wondered if he was the guy Feldman was talking about. End of story — or so she thought.

A month or so later the Chargers assigned her to attend a corporate NFL event in Palm Desert, where one of their major partners, GTE, was among the sponsors. It just so happened that Junior also would be there, though she didn't know it. This time he wasn't going to let her get away when he saw her. He invited Gina and a close friend who had accompanied her on the trip to join him that evening at Pompeii's, a local club. She was intrigued but still had her guard up, because, well, he was a big-time athlete and they were known to be womanizers. Junior sensed her reluctance, but he had a smile and a personality that could melt an iceberg.

When she showed up that evening, he quickly made his interest in her known—even though he had a girlfriend and a child back home. He'd hold her hand and put his arm around her shoulder. He looked into her eyes and made it seem as if they were the only two people in the room. He even got in a fight for her, knocking out a guy when he spotted him being too forward when he and DeBoer were separated.

Feldman immediately rushed Junior out of the club. He feared negative press could cause Junior to fall in the draft and told Gina and her friend to take off and meet them at his place later that evening. Once there, the group of five or six cooked tacos, listened to music, and hung out at the pool. Later, at around 11:00, Junior pulled DeBoer aside and told her he wanted to get to know her one-on-one. She wondered if he was running a game on her. The attention was flattering, but was it genuine?

"He was so interested in how I grew up—was I raised in a Christian home? How many brothers and sisters did I have?" she said. "They were questions that a man who's 21 years old and egocentric and self-centered wouldn't ask. I told him I grew up in a real conservative home, went to church every Sunday, Sunday school and catechism classes. It was something he could relate to.

"He said, 'Oh, do you remember this little song from Sunday school?' and he'd start singing it. It was the same songs I knew, but he sang them in Samoan from his Samoan church. I thought, *How sweet is this guy?* That was a good connection for me. I thought, *This guy comes from the same* [faith-based] *background as me, just a different culture.*"

They began spending more time together, and he won her heart with his kindness. For instance, he regularly tried to be available for dinner with her, but during one predraft stretch he was busy in Ocean-side for three straight days and didn't see her. On the fourth day he called her office and apologized, even though she said it wasn't neces-sary. He wanted to make it up to her, and so unbeknownst to her he booked dinner reservations and purchased a card and flowers.

"I thought, *Wow, what did I do to deserve this?*" she said. "It made it very easy for me to fall for him and to trust him and to feel secure and safe with him. He worked very hard to be a good boyfriend."

She tried to keep from falling too quickly, but she saw something different in Junior. He was a big-time athlete who wasn't full of himself. There was an almost childlike quality about him. He was an imposing mass of muscled humanity, but with a vulnerable, sensitive, innocent spirit. She likened him to Ferdinand the Bull, the physically imposing children's book figure who prefers smelling flowers to participating in bullfights. He was the type who hurt when she hurt, cried when she cried, laughed when she laughed. It was all fantastic except for one thing: Junior still was seeing Melissa Waldrop—and had yet to tell ei-ther of them that he was involved with the other.

This may have been an early example of Junior's willingness to lie and to manipulate situations. He would tell different things to differ-ent people, letting them know only what he wanted them to know. So even as he was wooing Gina in Palm Desert, making her feel as if she was the only woman in his life, he was calling Melissa and saying how much he missed her and Tyler and couldn't wait to get home.

Waldrop hadn't known anything about Junior when she transferred to Oceanside High from El Camino six weeks into her sophomore year. He spotted her one day on campus and immediately was smitten, telling a mutual friend that he thought she was pretty. After one foot-ball game they were part of a small group that went out together. The two quickly hit it off and were together from that point on. He even gave her a promise ring on their one-year anniversary. He was her first love, and she was his—or so she thought.

Her first indication that Junior might be cheating on her came after high school and college. In fact, it was on Tyler's first birthday. She and Junior had joined two other couples for a trip to Magic Mountain, an amusement park roughly three hours north of Oceanside. When they returned to the Newport Beach condo where Junior was staying, the phone rang. Junior allowed the answering machine to pick up. The only problem was that the volume was up, and Gina DeBoer was on the other end, saying she needed to speak to him before going to work the next morning.

Waldrop immediately questioned him about it. He gave her an explanation that calmed the situation but didn't allay her suspicions. The tension carried over to July 4, when she and Junior spent the day hanging out on the beach in Oceanside with friends and other couples. He had become distant, and an argument ensued. Finally, she broke off the relationship. "I hope your football keeps you warm at night," she said.

One month later Junior and Gina moved in together. The following spring he took her to a sunset dinner in La Jolla and ordered their favorite item on the menu: beef Wellington. She had barely taken two bites when he started asking if she was done. "Let's hurry up and get to dessert," he said.

If DeBoer thought Junior was acting strange, she found out why when the dessert menu arrived. On it, Junior had written: "Gina, I love you. Will you marry me?" She said yes through tears of joy, but the evening of bliss would quickly be followed by turmoil. Many of Junior's family members believed that Gina looked down on them and had no interest in being a part of an unrefined family. She had grown up in Danville, California, a well-to-do area that *Forbes* magazine listed as having a population of 52,078 in 2012, with an average home price of nearly $700,000. She went to a private Christian high school, grew up around mostly wealthy Caucasian kids, and attended San Diego State. Said another way, her upbringing was everything that Junior's was not, both financially and culturally.

Before proposing, Junior had a meeting to clear the air with his family at First Samoan Congregational Christian Church of Oceanside. The tension in the room was as thick as a San Francisco fog. Some siblings said they flat out were against the marriage; others merely expressed reservations. One cousin bluntly told him that he would not attend the ceremony if the marriage took place.

"Everyone has skeletons in their closet," Junior said, exasperated. "If you love me as much as you say you love me, you'll let the closet remain closed."

Despite Junior and Gina marrying in March 1992, the door to the closet never closed completely. It remained cracked throughout their marriage.

5

Dream Turned Nightmare

JUNIOR WANTED to play for the Chargers from the moment he turned pro. They had the fifth pick in the draft, and it had been his dream to suit up for the hometown team since he first began playing football as a seventh grader. At night he would lie on a bed and fantasize about wearing a helmet with lightning bolts on the side of it. Once, during his senior year in high school, the Pirates qualified for the CIF (California Interscholastic Federation)–San Diego championship football game, which would be played in the former San Diego Jack Murphy Stadium, home of the Chargers. On the eve of the game, the teams were granted a walk-through practice in the stadium.

During a lull, Junior walked up a long, narrow tunnel and down a short path to the Chargers' locker room. He entered with trepidation and excitement and stood in awe of the helmets and jerseys hanging in the dressing stalls. Chargers equipment manager Sid Brooks finally spotted the young man-child and asked what he was doing. Seau explained who he was and told Brooks that quarterback Dan Fouts was his idol. Then he asked if he could touch Fouts's helmet. Brooks agreed, leaving the youngster speechless for one of the few times in his life.

Four years later, Junior couldn't help but reflect on that moment and envision a jersey with his name hanging in the Chargers' locker room. But would he still be available when the Chargers were on the clock to make their selection?

With the first pick . . .

There was no suspense with this selection. The Indianapolis Colts acquired it from the Atlanta Falcons two days before the draft, then negotiated all night to come to terms with Illinois quarterback Jeff George on a six-year, $15 million deal that included a $3.5 million roster bonus. The Dallas Cowboys initially owned the pick; however, they had relinquished it the previous year when they selected quarterback Steve Walsh in the first round of the supplemental draft. With the number 1 pick out of the way, the real intrigue in the 1990 draft began.

With the second pick . . .

The New York Jets were on the clock and needed a pass rusher after finishing last in the league with 28 sacks. Rookie end/tackle Dennis Byrd led them with seven sacks as a situational player, but no one else had more than four and a half. If they needed confirmation of the value of a dynamic edge rusher, they could look to the New York Giants, a team with whom they shared the market.

In 1981 the Giants defense had gone from being the league's second-worst to its third-best because of rookie outside linebacker Lawrence Taylor, a ferocious, relentless talent who could take over games by himself. At the time there were a lot of similarities between Taylor and Junior in terms of size—Taylor was six-three and 237 pounds; Junior was six-three and 245 pounds—and also in terms of competitiveness. But the Jets elected to take Penn State running back Blair Thomas (who turned out to be a flop).

With the third pick . . .

The Seattle Seahawks, who acquired the selection in a trade with the New England Patriots, also needed pass rush help after finishing the previous season with 32 sacks, seventh-fewest in the league, but they felt good about their people on the edge. Linebacker Rufus Porter was coming off 10.5 sacks in only his second season, while starting just three games, and end Jacob Green had at least nine sacks in six straight seasons, including 10 or more in four of them.

The real concern for the Seahawks was on the interior. They were switching from a 3-4 scheme (three linemen, four linebackers) to a 4-3 scheme (four linemen, three linebackers) and didn't have a guy on the inside who could dominate. So they selected Miami defensive tackle Cortez Kennedy (who went on to become a Hall of Famer).

With the fourth pick . . .

This is where things got interesting. The Tampa Bay Bucs had made it fairly clear that they were going to select a linebacker, but would it be Junior Seau or Alabama All-America Keith McCants? New coach Ray Perkins had recruited McCants to Alabama and once said of him: "He plays like he is never out of the play. That is an intensity level I like." Still, McCants was not as dynamic a pass rusher as Junior, finishing with only four sacks in his final season. Junior had nearly that many in one game.

"We told all the teams that Junior wanted to go to San Diego and if you draft us we won't sign, so basically, go fuck off," said Feldman, his agent. "When the Bucs were on the clock, I hopped on the phone with them and made it real clear: My guy is San Diego–born and –bred, and that's where he wants to be."

The irony is that Junior repeatedly had preached that he wanted to be the first linebacker selected. He even tried to prop himself up by putting McCants down. The perfect scenario for him would have been for the Bucs to draft any position but linebacker, thereby allowing him not only to be the first player selected at his position but also to join the Chargers. He got one of his two wishes: the Bucs selected McCants (who played for three teams over six seasons and finished his career with just 13.5 sacks).

With the fifth pick . . .

It was time to see if the Chargers felt as strongly about Junior as he felt about them. Feldman previously had received assurances from general manager Bobby Beathard that San Diego would select Junior if he was available, but Beathard never expected Junior to be there. In fact, he didn't bring him in for a workout until late in the predraft process because he didn't want to waste his time or Junior's.

Billy Devaney, the team's director of player personnel, was extremely close to Beathard, and he kept asking: "What if you're wrong? What if Tampa Bay doesn't take him?" Finally Beathard relented and gave the okay to arrange a workout.

"Bud-dee," Junior said. "You're the only team [at the top of the draft] I haven't worked out for. I was wondering when you were going to get around to calling. What took you so long?"

The call actually may have been longer than the workout. Beathard,

Devaney, coach Dan Henning, defensive coordinator Ron Lynn, and a few others took Junior to the practice field beyond the parking lot at San Diego Jack Murphy Stadium. It was a welcome diversion for the coaches after being cooped up in their offices for draft preparation.

The group asked Junior to run laterally over four or five bags, then do a couple of pass drops. Junior knew they had questions about his ability to drop into coverage because he rarely did it in his final season with the Trojans. Linebackers coach Mike Haluchak was chosen to run the pass route, acting as if he were a running back. Junior jammed Haluchak so hard that Haluchak flew sideways, nearly lost his balance, and pulled a hamstring. End of drill. End of workout.

"Everyone looked at each other and said, 'What are we doing this for? This is stupid. June, that's enough. We've seen enough,'" Devaney recalled. "Even then, we never expected Tampa Bay to take McCants. When Ray Perkins did pick him, we started high-fiving and loving Ray Perkins. Our card with Junior's name on it went up to the podium so fast."

Junior pumped his fist and let out a loud "Woooo!" when he reached the draft stage in New York City. He was late in making his way out to greet Commissioner Paul Tagliabue onstage because he was on the phone with his parents. He repeated to them what he had said to them when he accepted a scholarship to USC: "Dad and Mom, I try hard for the family," his father recalled.

Junior was warmly welcomed by the community, but full acceptance by team members took time. Everyone knew he had been blessed with physical skills, but some viewed him as a hothead because of his on-field run-ins while at USC and his outburst in the Chargers' locker room during a predraft visit.

On that day Junior was receiving a tour of the locker room when defensive end Burt Grossman, the eighth pick in the draft the previous year, said loud enough for everyone to hear: "What's this guy doing here? Where's Keith McCants?"

"What did you say?" Junior responded.

"Junior wanted to fight me, right there in front of everybody," Grossman said. "It wasn't like he was hiding or trying to play cool. He wanted to come over to my locker, and they're pulling him away. I kept egging him on because I'm like, 'Ah, this fucker ain't getting drafted here. I'll never see him again.'"

Junior was a long way from his childhood, but the instinct to lash out when frustrated was still a major part of his personality. During one of his initial practices with the team, he and defensive end/outside linebacker Leslie O'Neal squared off because O'Neal did not huddle on his command.

"Junior had a chip on his shoulder when he first arrived because he had had the one breakout year at USC, but prior to that he hadn't really done anything," Grossman said. "He was insecure in the beginning, like he feared he wasn't good enough. You'd say something and he'd want to fight. It didn't matter who it was."

Because of that insecurity, he had trouble trusting people. When veteran linebackers Gary Plummer and Billy Ray Smith took him to lunch during an off-season minicamp, Junior thought it might be part of the rookie hazing he had heard so much about. He was wrong. Both Plummer and Smith had a mischievous side, but in this instance their intent was to make Junior feel like he was one of the guys after he made a favorable impression on them during workouts.

"The Super Samoan" is how Plummer referred to him in the *San Diego Evening Tribune*. "You can put two S's on his chest. Even if he screws up, he has so much speed he's going to make up for his mistake. To me, the most incredible thing is his catch-up speed. If he's beaten 1-on-1, he's got the kind of catch-up speed you see in someone like [Washington Redskins cornerback] Darrell Green but never in a linebacker. I've played next to a lot of guys, and that's all they were — guys. There's just no comparison. You don't expect to see that kind of athlete at that position."

Still, everyone knew it was critical for Junior to be signed and at training camp on day 1 because he had so much to learn. At USC, because he played outside linebacker — and defensive end in some passing situations — everything came at him from one side. That made it easier for him to read the offense and play with abandon. But the Chargers planned to use him at inside linebacker, where things would be coming at him from the left and the right. He had to be able to process information quickly; otherwise, he'd lose the half step or full step that often is the difference between making a big play and giving up a big play.

Beathard was scheduled to open contract talks with Feldman on the final day of the minicamp, which was held in May. "I've come here to be a football player," Junior said to the media at the end of the camp. "If the business part comes to a point where I have to sit out and lose experience, that would be distasteful. It's a big jump for me now, and I don't want to slip backward. If I come in late and I'm not in top shape, I shouldn't be the one to blame."

His words were foreboding. The prospect of a contract stalemate first materialized a week before training camp when McCants, selected one spot ahead of Junior, signed what was reported to be a five-year, $7.4 million contract that made him the highest-paid linebacker in football, with an average salary of $1.48 million a year. (The deal turned out to be for $1.2 million a year, just under what Taylor was making with the Giants.)

Feldman already had publicly stated that Junior wouldn't sign for less than McCants, and Beathard had privately let it be known that, based on the league's slotted salary scale for rookies, he was not going to pay the fifth pick more than the fourth pick. More troubling, Feldman disclosed that the sides were separated by an average of $1.1 million a year, a staggering figure with training camp so close to starting. Junior was seeking $1.8 million a year on a three-year deal, and the Chargers were offering $700,000 a year.

Beathard became irked when Feldman took the negotiations public. The agent argued that Junior deserved more than McCants because the two of them conspired to have Junior fall to the Chargers. He also argued that Junior was at least the second- or third-best player in the draft and would have gone higher if they hadn't scared off teams by saying that he'd refuse to sign with them if they drafted him.

Beathard wasn't moved, and he let it be known internally and publicly that he was not going to be bullied. He had the résumé to stand up to a popular first-round pick because he had been successful everywhere he had been, first as a scout or personnel man with the Chiefs and Dolphins, then as general manager of the Washington Redskins, who in the 1980s won two Super Bowls in three appearances under his direction. It's not often that the hiring of a general manager excites a fan base, but Beathard's arrival in 1990 marked the first time in six years that the Chargers' season-ticket sales increased. Fans trusted Beathard and believed in him. There was no chance of him being bullied, regardless of the fact that the Chargers had not been in the playoffs since 1982 and had finished last in the AFC West in three of the previous six years.

When training camp opened, Junior was nowhere to be found. A week passed ... then two ... then three. Junior joked when asked about his absence, telling the *San Diego Union:* "I've decided not to shave until I get signed, and right now I'm getting some hair on my face and looking kind of ugly. I told my agent when I start to braid it, we're in trouble."

An unintended consequence of his contract impasse was that Junior was losing the goodwill he had built up with his teammates during off-season workouts. They were starting to view him as a prima donna, not a happy-go-lucky kid eager to play for his hometown team. The day after making the joke about his beard, Junior and Feldman met with Beathard and proposed a contract that would average $1 less per year than McCants's deal. Beathard rejected it. Not only that, he said, because he considered the move "showmanship," he might not counter the proposal.

Things quickly went from bad to worse. Junior was so put off by Beathard's response that he, Feldman, and DeBoer flew to Cancun to get away. "I think it was our way to let everybody know we were going to do our own thing until we got a deal," Feldman recalled. "We body-surfed and hung out and danced and drank and did all sorts of crazy stuff. He was in great shape. He trained like crazy. So there was never a concern that he wouldn't be ready once a deal got done."

At that point Feldman was seeking an average of $1,199,999 a year while the Chargers were offering $850,000 — the same offer they had made 23 days earlier. By way of comparison, the number 5 pick in the previous year's draft, cornerback Deion Sanders, received an average of $1.1 million on a four-year deal from the Atlanta Falcons. Sanders had leverage, though. He was a talented baseball player, and the Falcons knew that Major League Baseball was a legitimate option for him.

Feldman began to feel the pressure a week later when three of his Chargers clients — Pro Bowl defensive end Lee Williams, who was seeking a new contract, cornerback Gill Byrd, and defensive tackle Joe Phillips — all fired him. "I don't think Feldman's in a position anymore to help me renegotiate my contract without a lot of hostility," Williams told the *San Diego Union*. "I think it's gotten to the point that there's bad blood between him and this organization. Quite frankly that leaves him a lame duck where I'm concerned. It took me about five days to make this decision because Steve never did me wrong; all he ever tried to do was help me. I hate doing it, but it's the big picture. He's not very well liked around here. I don't want to say there's a war going on between him and management, but I know I'm in the middle of something. And I don't want to suffer because of something that doesn't pertain to me."

On August 22, the Chargers made a take-it-or-leave-it offer and gave Junior until 5:00 PM the next day to accept it. The contract would pay him an average of $905,000 a year, roughly $300,000 a year less than what he was seeking. "If he rejects it, I guess it gives him a chance of not playing football this year," said Beathard, who added that the offer would be reduced daily if not accepted. Junior accepted it, after the deadline was pushed back 20 minutes because Feldman couldn't reach him by phone.

Happy days? Hardly.

Junior was supposed to report to camp that Sunday evening but never did. Feldman informed the team that Junior had missed his plane from Cancun, but he complicated matters by adding that his client was angry at the organization — not just about the way it handled negotiations, but also because Gina DeBoer had been fired during that time. The club said it was following an organizational policy that prohibited fraternization between staff and players; however, DeBoer said that she and Junior were told before the draft that their relationship would not be an issue with the team.

When Junior failed to report Sunday night, speculation mounted that he might renege on the oral agreement. Such talk irked an already irritated Beathard. "It's just a game, and we aren't going to play their game much longer," the GM told the media. "The best thing I can say is, if the guy is as unhappy as we read in the newspaper, then maybe it would be better if he didn't come here."

When Junior did arrive, he still was unsure about signing. He met with Beathard, then walked outside, where he, DeBoer, and Feldman took a stroll around the University of California–San Diego campus, home of the Chargers' training camp. Finally, after four days, three missed planes, and some figurative mending of emotional fences, he signed a five-year, $4.525 million deal.

There was no big press conference, no wide smiles as he sat before the cameras and scribbled his name at the bottom of a contract. Instead, an uneasy tension filled the air. He was happy to be a Charger, yet unhappy to be a Charger. The team received a two-game roster exemption to get him in shape, but the Chargers immediately announced their plans to play him that Saturday in Los Angeles in the preseason finale against the hated Raiders.

And so the real fireworks began.

6

"I've Got to Get Better"

LOS ANGELES SPORTING crowds are notorious for arriving late and leaving early, but September 1, 1990, was particularly dreadful. The LA Coliseum was virtually empty when the Raiders kicked off against the Chargers in the preseason finale for both teams. The announced crowd of only 25,071 barely registered inside the 90,000-seat stadium.

The sparse turnout wasn't completely shocking considering the game meant nothing in the standings and clubs typically limit or rest their starters in the final exhibition game, fearful of exposing them to injury. But this game was slightly different for two reasons: Junior would be making his unofficial pro debut after missing all of training camp because of a contract squabble, and he'd be doing it in the place he called home during his college career. Perhaps the public was wise to save its money because Junior, who took the field late in the first quarter, was ejected after only two plays for punching Raiders left guard Steve Wisniewski.

The strike came at the end of a play that's as fundamental to football as blocking and tackling: a tailback run into the middle of the line. Junior correctly diagnosed the play, but he charged too aggressively toward the line of scrimmage and missed the tackle on Napoleon McCallum, who gained six yards. Wisniewski, who had a reputation for playing to the whistle — and sometimes beyond — latched on to Junior from behind and pushed him toward the pile where McCallum had gone down.

"I knew something about Junior, being that he was at USC and we played in Los Angeles," Wisniewski recalled decades later. "I really wanted to push his buttons and test him. I got a chance to block him, and I just ran him, ran him, ran him and blocked him into the ground. Then I purposely sat on him to see how he would react. I was slow to get off of him, to test his mettle. Sure enough, he finally jumps off the ground and starts swinging. He made several rookie mistakes. One, he swung with his bare hand. Two, he connected with my helmet. And three, I just pointed to the ref and put my hands up, and they ejected him from the game. It was really comical. I couldn't believe what happened."

Wisniewski was the type of gritty, nasty player you loved to have on your team but hated to play against. The six-foot-four 305-pounder liked to annoy and antagonize opponents with his words as well as his actions, and Junior was an easy mark for him. Junior, intense and energetic by nature, was particularly amped for the game because he knew all eyes would be on him after missing all of training camp. He wanted to show the team, the community, and his family that they could be proud of him and depend on him, but instead, he reverted to the overly aggressive kid from the Oceanside Boys & Girls Club.

His ejection was greeted with an air of disbelief. It was a wasted opportunity for the Chargers, who wanted to see if he was far enough along to contribute in the season opener the following week. Two plays were not enough to make an informed decision.

Coach Dan Henning immediately called for trainer Keoki Kamau to escort Junior off the field and up the long tunnel to the visitors' locker room. Junior laughed at himself when Kamau arrived: "I didn't even break a sweat yet," he said. As they walked away Junior made a promise to himself regarding Wisniewski.

"I'm going to own that guy," he said. "By the time we're done, I'm going to work him over."

"I know, Bug," Kamau said. "Let's just get through this thing."

When they finally reached the locker room, Junior burst out laughing. "I can't believe I just got kicked out of the game," he said, incredulous.

"That was Junior," Kamau recalled. "He was angry at himself for being in that situation and not reacting the way he would've normally reacted. He should've taken the high road, but being a young athlete, coming into the league, full of energy—it was, let's go get this thing. Then he gets tangled up, and the next thing he's out of the game. We're walking off and he's like, 'This is unbelievable.'"

The expectation was that Junior would get a major dose of ribbing and snide remarks from veteran teammates who had gone through the meat grinder of training camp while he was on the beach in Mexico, but it didn't happen, at least not to the extent that anyone expected. If anything, Junior had earned the private respect of some of his teammates who viewed his intensity, athleticism, and physicality as the missing pieces on a defense looking to take the next step.

"We knew we had a 'player'—a guy who wasn't going to back down, who was going to set the tone," said cornerback Gill Byrd. "It didn't matter that he was a rookie—he was going to give it his all. We looked at that as something good, because he was setting a tone that we hadn't had in the past. We needed somebody in the linebacker corps who would elevate the play of everybody around him and play with a lot of passion. We had a lot of good, talented players on the team, but Junior had that something special about him as an athlete. It was different. You could see he genuinely loved and enjoyed the game. He was probably the best athlete with the biggest motor I had ever seen to that point. You knew greatness was there."

There was no need for teammates to get on Junior because privately he was chastising himself. Feldman, his agent, had prescheduled a family get-together for the Seaus that night after the game. It would be a chance for everybody to celebrate the signing of his contract because there had been no opportunity to do that earlier. Junior had reported immediately to the practice field after coming to contract terms, so there had been no time to visit with family.

The mood at Feldman's Newport Beach condo was far from festive. With time to think on what had transpired, Junior was even more embarrassed.

"He was devastated," said Gina. "He said to me, 'G, there's so much I have to do to prove myself, to gain the confidence of our city and my teammates and coaches. I have so far to go.' I was like, 'It's okay. It happened. You'll be fine.' I was trying to be positive and encouraging, but I didn't know him yet as an athlete. I didn't know how seriously he took this. He knew his talent, and he had something to prove to his team-mates, the coaches, and his family. He wanted to make his father and mother and the Seau family proud again."

It would not happen the next week because not only did Junior not start the season opener — Cedric Figaro did — but he also was completely ineffective as San Diego was outscored 10–0 in the fourth quarter and lost 17–14 in Dallas. Junior, who replaced Figaro after the fifth snap and played the rest of the way in nonpassing situations, had only four tackles and frequently was out of position. Not only that, but his 15-yard penalty for spearing contributed to a Cowboys drive that produced the decisive field goal.

When he got home and spoke to Gina, he kept repeating himself: "I've got to get better. I've got to get better."

That chance would come the following Sunday in the home opener, against the Bengals. Henning moved him into the starting lineup in hopes of gaining a spark, but the crowd of 49,679 — which was well below capacity — gave him a chilly reception during player introductions. A noticeable number of fans booed when he ran onto the field after his name was called.

If the greeting was embarrassing for Junior, it was awkward and uncomfortable for Gina. Her relationship with Junior had become so serious that her mother had flown in from northern California for the game. She had yet to meet Junior, and now she was forced to balance the good things she had heard from her daughter with the boos she was hearing from the crowd.

Some fans were upset that Junior had acted so put off after signing his multimillion-dollar contract. He had agreed to terms on a Thursday but didn't show up for work until the following Monday, a day after he was supposed to report. Then he spent 90 minutes walking around the UC San Diego campus with his agent and girlfriend, trying to decide if he really wanted to sign the deal. "It's not the best, but we're willing to deal," he told reporters.

In the local newspaper, one fan spoke for countless others in a letter to the editor. "I am tired of reading about how unhappy Junior is," the fan wrote. "How many kids just out of college, with a degree, receive a salary of $1 million for their first job? Junior should be grateful that anyone is willing to pay him such a ridiculous salary. If this crybaby continues to pout, the Chargers should point him to the nearest fast-food employment office."

Another wrote: "I'm currently in a homeless shelter, and I bet many people here who attend chapel services have acquired a more peaceful and invigorating spirit than you, Junior! Feeling empty these days, Junior? Oh, shucks. C'mon, meet me over coffee sometime at the shelter. I'll even buy, Mr. Millionaire. Anything, if only I can encourage you to 'wake up and smell the coffee,' to 'get a life' of ultimate value and contentment."

Junior had unintentionally become the symbol of the greedy young athlete who wants everything without doing anything. So fans booed whenever his name was called during the home opener. Their behavior clearly impacted him. He finished with only five tackles as the Chargers surrendered the final 14 points and lost 21–16.

"That hurt," he said two years later of the treatment. "I've never had an experience like that, and it's something I don't want to go through again. That gave me a sense of the real world. I want to gain respect, and that was humbling. I think it helped me out as a player and as a person."

It should have been a happy time for him: he was playing in his hometown, fulfilling his boyhood dream, and providing for his family. But Junior was miserable. The Chargers moved him to inside linebacker, and he hated it. He wanted to be back on the outside, where he had a clearer path to the quarterback. He expressed his feelings to the coaches and management, but they kept him on the inside. His emotions fluctuated between anger and self-pity. He had had so much success at outside linebacker in college that he couldn't understand why the Chargers wouldn't set him on the edge of the offensive formation and allow him to rush the passer, as he had done at USC. That was how he was accustomed to making plays. It allowed him to be special.

In the Chargers' defensive scheme he felt like just another guy. Worse, his penalties were costing them games. Besides the one against Dallas, there was another for unsportsmanlike conduct in a 17–7 loss to Houston. Football was no longer fun for him, and it showed in his personality. He didn't smile as much, and he kept his playful side behind a curtain of apparent indifference.

He would show up on time for work, then leave as soon as practice ended. It was the first time in his life he didn't want to be in the locker room or to hang out with teammates. His parents' home offered little refuge because his father was upset with the negative attention Junior had been receiving. Papa Seau felt that the family name had been tainted by the bad press.

"Junior was born with the Seau name, and Dad looked up to Junior to bring his name up," said Mary, his oldest sister. "Dad was upset after [the negative reaction to the contract impasse]. I asked him, 'Why are you getting mad at his glory? He worked for it, you didn't.' Dad said, 'That's so embarrassing.'"

The same sense of disappointment and shame that Junior felt after failing to receive a qualifying score on his SAT in high school came rushing back. A high point in his life had become a low point—again. He also raised eyebrows with his lack of humility, such as when he repeatedly spoke poorly of McCants, the linebacker selected ahead of him in the draft.

In April, before the draft, Junior tried to prop up himself by putting down McCants. "Everyone made him out to be Superman, but then they saw he didn't fly," he said. "Everyone wanted to give him his cake, but then he ate it. He came in fat because he thinks it's easy. He didn't excite me at all."

The comment was out of character, but some wrote it off as predraft bluster. Six months later, however, before the Bucs and McCants were set to travel to San Diego to play the Chargers, Junior was still targeting the former Alabama star.

"I still hold it against him that he was portrayed as the linebacker of the year [coming out of college]," Junior said. ". . . I wish he played running back, then I would be satisfied because he would be getting hit hard and, hopefully, it would be by me . . . You can't compare sacks between him and me in college either, because there's no comparison. He would do me a favor if he brought up sacks because then I would say: 'Line both of us outside and see who gets to the quarterback first.'

What he has to understand is that I'm a competitor. He was portrayed as the best and I went after him. Deep inside me—and I think a lot of people will vouch for me—I beat him. I beat him on the field and I beat him in the Combine workouts."

Junior then noted that he was starting as a rookie while McCants was playing as a reserve. Junior ranked fifth on the team in tackles while McCants was 15th.

"He's got his money, but, then again, I can say, 'I'm playing and you aren't,'" Junior said. "I mean, it would hurt me to get $1.2 million [a year] and sit on the bench the rest of the year or not be considered as a player they need to have in the defensive lineup to help the team win. I love responsibility, but if I didn't have the responsibility and still got the money it would be hell."

The comments angered Papa Seau. Humility is a hallmark of traditional Polynesian culture, and Junior's self-centered comments brought more embarrassment to the family.

"The one person I look up to is my dad," Junior told the *Los Angeles Times* the following year. "He's my hero. Dad was mad because of all the perceptions out there about me. The only way people knew me was through the newspapers and the media, and I didn't click. He was giving me the silent treatment and the glare. My dad had sacrificed everything, and I was screwing up. He took it hard when we lost. In high school, if we lost a game, my dad wouldn't give me any lunch money. If the team lost, it was because I hadn't done enough. That has been [ingrained] in me."

His image was taking a beating in the locker room as well as in public. Some teammates felt that the front office and the coaches were coddling and protecting Junior. It was customary on the day after games to review the film en masse or as a defense, but that didn't always happen if Junior had a bad performance or a major screwup. Whether one thing had something to do with the other is unknown, but the reality was that some teammates wondered privately about it.

"We're not meeting as a team to go over the film? Junior must have fucked up," recalled Leslie O'Neal. "It got to the point where there were a couple of games the coaches didn't grade the film. The way the organization is, they want their guy to be the guy that you think about when people say 'San Diego Chargers.' That's what they wanted with Junior, and they did everything in their power to make that happen. Billy Ray Smith was the guy, then I came in and I became the guy—but they didn't want me to be that guy anymore, maybe because of my [unfiltered] personality. I felt I got paid to make plays; if you asked me a question, I was going to give you my answer, unlike some people who say what the organization wants to hear. They did everything in their power to put Junior in that role, and once he had that role things got out of control."

Junior started the final 15 games as a rookie and finished second on the team with 85 tackles. They were decent numbers for most rookies, but not for him. He measured himself by impact plays, and he ended the year with no interceptions, no forced fumbles, no fumble recoveries, and only one sack, which came in the final game. It was unacceptable to him that safety Martin Bayless had three times as many sacks as him and reserve cornerback Donnie Elder was only a half-sack behind him.

"You could see from day 1 that he had the total package, because people built like that don't run like that," Burt Grossman said. "But you could see he had no idea how to handle what he had. He was like a 16-year-old with a Formula 1 car. It was just veering off and crashing into everything. When he got it under control, that's when he became really special."

7

The Turnaround

THE KNOCK at the front door on a March morning in 1991 surprised Junior. He wasn't expecting anyone, and his home in the hills of Mount Helix was not someplace you stumbled upon. You got there because you meant to get there.

Junior gazed through the peephole and saw linebackers coach Mike Haluchak standing on the porch. *What's he doing here?* he thought to himself.

Junior was still trying to escape the turmoil of a difficult rookie season. The contract standoff, the booing from the fans, the fighting with coaches over which position he should play, the silence from his father, the feeling that he had been nondescript on the field — the memories still lingered in his mind. Now a reminder of all those things was knocking at his door.

Junior opened the door and greeted Haluchak with a half-smile and a confused expression. The coach asked for a few minutes to talk. He had something he wanted to say, and it needed to be said in person, not over the phone. Haluchak thought Junior had the potential to be a transcendent talent, but he knew it would never happen as long as the temperamental star kept fighting the coaches — and himself, to some extent. When the two sat down, Haluchak was succinct. "You have the ability to be one of the all-time greats," he said.

Junior bristled inside. *Why are you bringing this here?* he thought to himself.

It was clear that the coaches could see something in Junior that

Junior could not—or would not—see in himself. "Give it a chance," Haluchak said.

If the coach wanted a commitment, he didn't get it. There was an awkward silence when the two parted company. Haluchak wondered if he had gotten through to his prized pupil, but there was no way to know.

When the door closed, Junior held up a figurative mirror. The person staring back at him was someone he didn't recognize or like. Never beholden to material things, he was now consumed by them. The young boy who grew up sleeping on the garage floor now had a lavish new $1.14 million home for himself in the hills and a newly constructed home for his parents. The youngster who was raised to be humble now had a Mercedes-Benz coupe with SAY-55-OW vanity plates. The innocent child who was content eating his mama's home-cooked meals was now feasting at the finest dining establishments, places he'd never even thought about entering in his youth.

Junior had allowed himself to be corrupted by fame and fortune. He realized the only thing worse than being poor was being rich—well, rich with no priorities. "You never realize how much a million dollars is until it's slapped on you," he told the *Los Angeles Times*. "So many people are chipping at you, and they all sound good. I lost myself. The money began to run me. Being at USC, you see all that money flowing around, and I was never a part of it. Now I had my chance. I was trying to be Ralph Lauren instead of Junior Seau from the slums in Ocean-side. I had to get back on track."

For Junior, that meant working. He committed himself to training and studying. Although still frustrated with the team's plan to use him at inside linebacker, he wasn't going to let it be quicksand beneath his feet. He had to keep moving forward. He had to make plays. *I have to beat myself.*

The Chargers opened the season with a 26–20 loss at Pittsburgh. Junior was unspectacular, finishing with just three tackles and no sacks. He was marginally better the next week in a 34–14 loss to San Francisco, recording five tackles and no sacks. Those performances weighed on him. He was supposed to be a difference-maker, yet he had made no significant impact on either game. That began to change in week 3, though.

The Atlanta Falcons, leading by three with just under three minutes to play, faced a third-and-1 from their 44. A first down would all but enable them to run out the clock and preserve a 13–10 victory. Chris Miller took the snap and handed the ball to five-foot-seven, 201-pound running back Steve Broussard, who needed only 36 inches for the first down. He not only failed to get the first down but was dropped for a one-yard loss when Junior burst through for the tackle.

It was the first time in the young season that Junior had made a potentially game-altering play; the stop helped put the Chargers in a good-enough field position to drive for a potential tying field goal. John Carney's 47-yard attempt sailed wide left, but the silver lining was that Junior had flashed the impact capabilities for which he was drafted. He continued to flash the following week against the Broncos, with nine tackles and another sack. He added this third sack in as many games and five tackles the next week against the Chiefs.

Each big play was followed by a powerful fist thrust that came to be known as his signature move. It represented a metaphoric blow against not only the opponent but also anyone who doubted whether he'd reach his full potential. Junior had finally bought into the system, and the result was a breakout season that earned him the first of 12 trips to the Pro Bowl. His 111 tackles were 50 more than he tallied as a rookie, and his seven sacks were six more than he had his first year. Haluchak beamed like a proud parent at the end of the season.

"One thousand percent," he said of the improvement from year 1 to year 2.

Junior's attitude was also evolving. On most Friday nights he could be found on the sideline of high school football games in Oceanside, where he was able to stay connected to his roots. "See, this is what football is all about here," he said while watching Oceanside battle El Camino in early November 1991. "It's fun. It's one of those things where you just come out and play. It's free, really. It's pride, and it's the honor of representing your school, of representing somebody. I can remember when I had nothing. I was just one of the guys that played three sports and didn't really know anything about the professional level or the college level. I just cared about getting on the field and playing. I don't want to lose contact [with that]."

Sullen and often unapproachable as a rookie, he was starting to resemble once again the youngster who had befriended social outcasts in the neighborhood. He smiled more and socialized with kids and fans in his old neighborhood. On this night he even agreed to phone a boy who had cerebral palsy. Nine months later, another youngster from North County got an unexpected call from Junior. It was Chris Fore, an offensive lineman at Fallbrook High who had suffered serious injuries to his left leg and foot in a car accident.

Fore was the passenger in a car driven by teammate Justin Patterson, a 16-year-old who had just gotten his driver's license in the mail that day. He was driving Fore and another friend home when he lost control of his SUV and slammed into a telephone pole. Sadly, Patterson died in the accident. Fore, who would have seven surgeries over the next two years, was depressed about losing a close friend and not being able to participate on the football team.

When Junior heard the story, he arranged to phone Fore. He would not allow the youngster to drown in a pool of self-pity. He spoke to him energetically, telling him to find a way to be a part of the team. Fore went on to earn the DeNormandie Award, given annually to the most inspirational member of the team.

"That would've never happened without Junior's encouragement and challenge," Fore said. "Junior inspired me. He told me of an injury he had at USC . . . He missed a considerable part of the season. He said he felt bad for himself at first, which is how I was feeling. Then he went on to tell me how selfish that really was, because the team was moving on with or without him. 'With or without me, there was going to be a Trojan football season. Hate to say it like this, Chris, but with or without you, there will be a Warrior football season. You have to find a way to be a part of that.'

"Nobody had challenged me that way before," Fore continued. "Here I was, going to be in a wheelchair and on crutches for the entire football season, and nobody had told me that I could still be involved until Junior did that night. I gained a new perspective. Football meant so much to me, so I started to look at it through a new set of eyes that night. I saw my teammates as people that I could still motivate, encourage, and challenge. So I did just that. The next year, as a senior, I coached the freshman team. I've now coached 14 years, eight as a head football coach. I don't know that I would've ever realized this path my life has been on without that call and a subsequent visit to Chargers practice where Junior further encouraged me in that regard."

The compassionate side of Junior was what heightened the community's respect for him. No matter how high he climbed, his heart was always with those not as fortunate. In the spring of 1992, he and Gina founded the Junior Seau Foundation to help at-risk kids in San Diego County. His particular focus was his hometown of Oceanside, where he was barely five years removed from being one of those kids facing the decision to do the right thing or turn down a destructive path. In many ways the early years of the foundation were like Junior's early years in the league, in that he knew what he wanted the foundation to accomplish but it would take years to achieve it because he lacked the training and experience. There was so much for him to learn.

On the field, he was quickly learning about the business of football. After three straight 6-10 seasons, the Chargers finished 4-12 and in last place in 1991. Beathard had great respect and affection for coach Dan Henning, but he also had an impatient owner in Alex Spanos, who wanted change. So Henning was fired and replaced by Bobby Ross, a no-nonsense Southerner who had won a share of the national championship two years earlier at Georgia Tech.

The hiring was far from inspiring to fans and team members. Ross was a slightly built former military man with a country drawl and a noticeable lisp. He had no NFL experience, was coming off an 8-5 season, and was just 31-26-1 overall at the university. He was hired to make things better, but the situation initially deteriorated. The Chargers lost his pro debut 24-10 to the visiting Chiefs, then traveled to Denver and were beaten 21-13. In week 3 they surrendered 16 fourth-quarter points to the Steelers and lost 23-6 at home, then saw the bottom fall out the following week in a 27-0 defeat at Houston.

As if 0-4 weren't bad enough, the Chargers had been outscored 61-10 after halftime. The second-half failures pointed directly at the coaching staff and its failure to make successful adjustments. Complicating matters was the struggle of quarterback Stan Humphries, who had been acquired in a preseason trade with Washington after starter John Friesz sustained a season-ending knee injury in the exhibition opener. Through four games (three of them starts), Humphries had thrown eight interceptions and only one touchdown.

"We are 0-4, and it's no fun," Junior wrote in his personal journal.

Things are now starting to stir up in the locker room and outside with the media. When you or any team starts with this type of record, you start to question each other [players and coaches]. Things are good for me individually, but we are losing as a team. It's funny because going into the games I am determined to dominate the field. But when we lose I wish I could trade my performance for a win. Today I was told that there are some changes in the defensive scheme. Last week we played Houston and the pressure I put on Warren [Moon] was there and more. But there were some breakdowns on linebacker support, so this week we are placing me on the running back instead of rushing the QB. My frustration derives from working hard and doing a job that is being praised by viewers, and now being asked to do more. We have three other linebackers who can cover running backs just as good, but there has been some controversial moves with players on the team — i.e., ever since I've moved down to rush the passer my buddy Burt [Grossman, a defensive end] has been vocal about his feelings about the move. Burt had to move inside of the guard so I could rush on the outside. He's a good player, but we all know why I was put out there: speed.

Another issue is that [defensive end] Chris Mims, a first-round pick, has not played and the pressure on upper management has increased since the losing streak started. In order to take some heat off the upper office, Chris has to be plugged in; and for us to do that I have to move to linebacker, Burt moves to end and Chris goes to tackle. How convenient. My frustration is the lack of commitment to winning. If the front office is crying because they are looking bad, then join the club. If our coaches are sick of someone complaining about his position that they're willing to sacrifice a great performance for happiness or peace of mind, then we have a bigger problem than just not winning. I will say one thing, and that is that they are testing my character. I could be taking this too

strong or personal, but when your coach doesn't know how to explain
the situation or give me an understanding, you have to confront it.

Junior wasn't the only one to see the glass as half-empty. *Los Angeles Times* beat writer T. J. Simers wrote that Ross was "in over his head," which prompted an angry Ross to request a private meeting between the two. Overall, the situation was so bad that Ross was introduced at a boosters' luncheon as "the director of *The Laurel and Hardy Show.*" For someone as prideful as Ross, the comment was hard to take. But he also was a realist, and he knew his team hadn't played well enough to earn anyone's respect.

Fortunately for the Chargers, the NFL season is a marathon and not a sprint. They responded by beating the Seahawks 17–6 in their next game, holding an opponent without a touchdown for the first time all year. Then, after returning from the bye week, they outscored the Colts 17–0 in the fourth quarter for a 34–14 victory. They followed that with a win over the Broncos, then another one over the Colts. They were back to .500 and no longer the butt of their boosters' jokes.

When the streak was broken the following week against the Chiefs, no one took the 16–14 loss harder than Junior. The Chargers led by one with just over two minutes to play and Kansas City in possession of the ball at midfield. They had held the Chiefs to four straight three-and-outs, but on first down Junior blew a coverage that resulted in a 25-yard catch by Willie Davis. Three rushes later, Nick Lowery kicked the decisive 36-yard field goal.

Junior wrote in his journal:

I'm hurting really bad for my team, but also disappointed with my lack of concentration level during that ONE play. I've learned how lucky I am to have another opportunity. Note to self: Never take a down off. It may be the most important play of the day.

Putting his feelings on paper proved unfulfilling, so the day after the defeat Junior stood before the team and apologized for his mistake.

"He broke down and started crying," Grossman said. "That was the type of person he was. You just don't see in this day of egos and grown men somebody that emotional. You don't see people in the NFL with absolute, stripped-bare remorse over football. I don't see people when their own family members die have that much remorse and that much pain that they let somebody down. I remember thinking at the time, *It's just a football game.* I had already forgotten about the play by that time, to be honest with you. This guy had the reaction that, I was drunk and ran over your son. It was like he was talking to the parents of a kid he had just killed. That's the kind of emotion he had.

"It was amazing to see because at the start of his career he had all these insecurities about, 'Am I good enough? Can I play?' People thought he was brash. But now he had morphed into someone who was beloved. That emotion was as real as it gets."

Two days later Junior's growing maturity and comfort level with the organization came through in another journal entry. Grossman was expected to miss the game because of injury, and the coaches phoned Junior on his off day to see if he'd be comfortable playing along the line in passing situations. He wrote:

> *I told them that I'll play D-line but don't take me out of the 46 [blitz] package entirely. Their response was respectable in this sense: The coaches showed some respect toward me as a person and player. "It would be easier if we had two of you," they said.*

He continued:

> *I've grown into a student of the game of football. Playing with rare ability and talent, I've gotten through hard times in the past. But this year I truly understand our schemes, and I'm aware of what my teammates are doing. Before, my reads were derived from film work and I would do anything to get the man with the ball — even though my responsibility was to do something else. The difference this year is that I don't have to put out or waste energy because I know our schemes. I can account for my players and know where to stunt them, then run to the void area.*

As he sat there writing, Junior reflected on his struggles over the previous two seasons, when he fought with the coaches about being moved from outside linebacker to inside linebacker.

When I asked God, "Why are they taking me out of the pass rush and putting me in the secondary," HE answered my question. I'm thankful. I've learned so much about the structure of the defense and offenses, but the exciting part of all this is, I have a lot to learn and do before I can say I DID IT!

Linebackers coach John Fox learned this early in the 1992 season, his first with the Chargers. Fox can't recall the specific month, the exact opponent, or the precise point in the game. But he'll never forget the play.

Junior was supposed to drop into pass coverage and protect against a hook route or crossing route. But at the snap of the ball he took off toward the line of scrimmage, burst past blockers, and sacked the quarterback to force a punt.

When he got to the sideline, Fox greeted him with one question: "What in the world are you doing?"

"Making a play, Coach," Junior said.

"He was serious as a heart attack," Fox said. "He didn't make an excuse or give me some bullshit. He just said, 'Making a play, Coach.' There are guys who can diagram it on the board and tell you what everybody is doing. Then there are guys that probably can't put it on the board like that, probably can't explain it like that, but instinctively they know what to do. Junior was both."

The visitors' coaching booth in Oakland is separated from the press box by a relatively thin piece of Plexiglas and vertical blinds. When the Chargers played there in Junior's first two seasons, the media regularly heard defensive coordinator Ron Lynn go ballistic during games. "Junior! Junior! What the fuck are you doing?!" he'd yell at the start of a play. "What the fuck are you doing?! Junior! Junior!"

In the next breath they'd hear: "Fucking great play, Junior! You lucky son of a bitch."

It was like the basketball coach who yells, "NO! NO! NO!" as a shooter launches a three-pointer, then claps and quietly says, "Yes!" when the ball swishes through the net.

Junior trusted his instincts and his ability to make a play more than he did the call being sent in by the coordinator. It's no coincidence that the Chargers started to come into their own just as Junior was doing the same on the field. Despite the horrible start to the 1992 season, they won 11 of their final 12 games — including the last seven — to claim their first division title in a nonstrike year since 1981. They also became the only team in league history to reach the playoffs after a 0-4 start, ending the AFC's longest active postseason drought at nine years.

Junior's intensity helped fuel the streak just as his playfulness helped create a family atmosphere in the locker room. He loved to prank teammates, with Grossman being a primary target. The two had become close friends since their tense initial meeting before Junior was drafted. They roomed together for two years on the road and in training camp, and Junior was the best man at two of Grossman's weddings. Very little was off-limits between the two, so Junior didn't hesitate to pounce when a gotcha opportunity presented itself. One such occasion involved Grossman's mail.

"They used to intercept Burt's mail, and in doing so they ran across one fan who had a big crush on Burt," a teammate recalled. "Junior's idea was to respond back as if they were Burt. Then each week they'd intercept the mail to see what the fan's response was. Well, Burt had no idea that any of this was going on. The reason this was so wild was because the person who was writing Burt and who had the crush on Burt was a male.

"One day the guys decided to take it to the next level and set up a meeting with the fan, because the fan was like, 'Hey, I want to meet you.' They were like, 'Sure, no problem. I'll be at this place at this time. I look forward to meeting you.' The guy actually showed up and was like, 'Burt, Burt . . . ,' and Burt's wondering what the hell was going on. The fan was like, 'Thanks for taking the time to meet me,' and Burt was like, 'What the hell are you talking about? I don't even know you!'"

Burt soon figured out that he was the victim of a practical joke, and that the likely culprit was Junior. He was angry, to say the least.

"The guy was gay, and Junior sent him flowers in my name," Grossman recalled. "Junior was a character."

Grossman tried to pull the same prank on Junior, but it failed to work because the element of surprise was gone.

"He got this tag that he was dumb because of Prop 48 and stuff," Grossman said. "But Junior was probably one of the smartest people I've met. Not so much book-wise or even your vocabulary, because he was notorious for butchering words and using them in the wrong place. But in terms of common sense and reading people, he was a really, really bright guy. So pulling practical jokes on him or things like that, it was really hard to get over on him."

Junior's fingerprints were prominent on the late-season win streak as well as the joke on Grossman. With the Chargers trailing early in the fourth quarter against the Cardinals, he forced a fumble that San Diego quickly converted into the decisive touchdown in a 27–21 victory. Two weeks later, with a chance to claim the division title with a win, the Chargers seemed to be prepared to make things more interesting than necessary.

They were dominant through two-plus quarters, building a 23–0 lead that seemed insurmountable. Then Los Angeles replaced ineffective quarterback Jay Schroeder with veteran Vince Evans, who sparked the Raiders with a 21-yard touchdown pass. The Chargers weren't in danger, but when they punted on their next series, momentum appeared to be shifting sides. Another touchdown could have made for an interesting fourth quarter. But Junior would have none of it. On third-and-18 from the Los Angeles 16-yard line, he picked off Evans's pass and returned it 29 yards to the Raiders' 3-yard line. Two plays later Eric Bieniemy rushed in from the 1, and the Chargers went on to win 36–14.

Junior, who also had a sack in the game, was ecstatic. He was playing well, and the team was on its way to the playoffs for the first time in a decade. It was no coincidence that the defense began to excel at the same time that he began to dominate. The unit allowed only one opponent to score more than 14 points over the final seven games, after allowing six teams to do so in the first nine games.

With every big play he made, with every powerful thrust of his right fist into the air, San Diego fans fell harder and deeper for him. The franchise always had been known for its offensive stars—from Paul Lowe to Lance Alworth to Dan Fouts to Kellen Winslow to John Jefferson to Chuck Muncie. But Junior was the first defensive player with the skill and flair to stand on hallowed ground with them. Despite being in only his third season, he had become the face of the franchise as it prepared to meet Kansas City in the opening round of the playoffs.

The Chiefs had beaten them twice during the season and six straight times overall, so to outsiders it appeared to be a horrible matchup. The Chargers saw it differently. To them, it was the perfect opportunity for payback. The front office also had added incentive entering the game. Beathard didn't like it that his KC counterpart, Carl Peterson, had privately accused San Diego of signing former Chiefs tight end Alfred Pupunu early that year for strategic purposes. The story was that the Chargers wanted Pupunu only so he could debrief them on KC's offensive playbook and adjustment calls.

The buildup to the game was intense, just like the hitting in the first half. The defenses dominated, with the Chargers gaining just 130 net yards of offense to the Chiefs' 122. The Chargers punted three times, the Chiefs four times. There was the sense that the first big play could decide the outcome, which proved to be true when halfback Marion Butts burst free for a 54-yard touchdown on the Chargers' second possession of the third quarter. It would be all the points San Diego needed in a 17–0 victory that represented the first postseason shutout in franchise history (13 games).

As the final moments ticked off the clock in San Diego Jack Murphy Stadium, the PA system blared the disco-themed "San Diego Super Chargers" song, which hadn't been heard since the glory days of Don Coryell, Dan Fouts, Kellen Winslow, Charlie Joiner, and Chuck Muncie. The chorus spoke to the team's performance that day, as Leslie O'Neal, the Chargers' standout edge rusher, finished with two sacks and an interception and led a defense that forced three turnovers and had seven sacks — six in the second half.

The next week they learned what it was like to be on the other end of the whipping stick as the Dolphins spanked them 31–0 in Miami. The Chargers were never in the game. Humphries, who had been so solid late in the year, threw three interceptions, each of which set up a Dolphins touchdown. Defensively, the Chargers surrendered three second-quarter scoring passes to Dan Marino and, overall, allowed the Dolphins to run for 157 yards and a touchdown on 40 carries.

"For the city of San Diego, we wish we could take [a victory] back home," Junior said afterward. "But it's obviously not meant to be."

Later, he took out a full-page ad in the local newspaper to thank the fans, whom he had won over with his aggressive playmaking on the field and his gentle compassion off it.

8

Super Fold

WEEK 1 is special in the NFL. Each team is undefeated, there's an air of optimism in every locker room, and winter has yet to put a death grip on parts of the Midwest and Northeast. Hope can fade quickly, though, especially when you're coming off a disappointing season, are picked to finish last in the division, and fall behind 17–0 in the first quarter of the season opener — in a city where you've lost seven straight games against an opponent who has won five consecutive home openers.

In 1994 that was precisely the situation the Chargers found themselves in with 10 minutes, 34 seconds, gone at Denver. They went three-and-out on their first offensive possession, shanked the ensuing punt for a net gain of only 13 yards, then surrendered a 50-yard touchdown pass to trail by a touchdown.

They ran four plays on their second possession, punted, and then allowed completions of 10, 18, and 13 yards that were complemented by runs of 17 and 22 yards, the last for another touchdown. As if that weren't bad enough, they fumbled away the ensuing kickoff to set up a field goal.

Suddenly their 11 wins in 12 games and playoff victory over the Chiefs in 1992 became an even more distant memory (after a lackluster 1993 season), replaced by images of the 10 consecutive years without a winning record in nonstrike seasons. *Same old Chargers,* many fans thought to themselves.

That's when Humphries, their quarterback with the strong right arm and portly middle, found tight end Alfred Pupunu for a 22-yard touchdown at the end of the first quarter . . . and Shawn Jefferson for a 47-yard score early in the second quarter . . . and Mark Seay for a 29-yard touchdown one minute and 45 seconds later. A rout was suddenly a game, and when safety Stanley Richard returned an interception 99 yards on the final play of the half, San Diego had its first lead, 27–24.

Even when quarterback John Elway, owner of 34 come-from-behind wins in the fourth quarter or overtime, drove the Broncos from their 25-yard line to the San Diego 3 with 43 seconds to play and Denver trailing by three, the Chargers refused to blink. Elway rolled right, cocked his cannonlike right arm, and then . . . inexplicably . . . had the ball slip from his hand as he prepared to release it.

Junior snatched it from the nighttime air to preserve a 37–34 victory that would set the tone for the greatest season in franchise history. When the play was over, he high-kneed his way to the San Diego sideline, where he hugged and high-fived his teammates.

"I saw a bright star," said Junior, who had a game-high 14 tackles. "It was a ball twirling in the air."

The surrealism of that night didn't end with the game. It continued for six weeks, with the Chargers knocking down anyone who got in their way. Broncos . . . Bengals . . . Seahawks . . . Raiders . . . Chiefs . . . Saints — they all fell. The Chargers' 6-0 record matched the best start in franchise history, equaling the mark that was set in 1961, the team's first year in San Diego after one season in Los Angeles.

The most impressive thing about their success was their ability to triumph in different ways. They won with offense, scoring 37 against the Broncos, 36 against the Saints, and 35 against the Seahawks. They won with defense, limiting Cincinnati and Seattle to 10 points each and Kansas City to six. They also won with special teams, beating the Raiders on a 33-yard field goal with two seconds to play.

The idea of Junior raising his game seemed ridiculous entering the season, in that he had gone to the previous three Pro Bowls. But as his understanding of how opponents might attack him caught up to his physical abilities he became a nearly unstoppable force. He had 13 tackles and a forced fumble against the Saints, 12 tackles against the Chiefs, 10 tackles and a sack against the Seahawks, eight tackles and two sacks against the Raiders, and six tackles and a pass defensed against the Bengals. His fist pump into the air seemed to be on a perpetual loop.

"He's a hellacious football player," Eagles defensive end Greg Townsend said that year. "They talk about Lawrence Taylor and Mike Singletary, but he's like both of them put into one."

Part of it had to do with being in a good place emotionally, something that couldn't be said in 1993 when Junior was confronted with serious family issues. His wife Gina was put on bed rest owing to complications with her pregnancy; their daughter Sydney was subsequently placed in a neonatal intensive care unit after being born seven weeks early. And Tony, his 16-year-old brother who was the baby of the bunch, pleaded guilty to attempted murder charges in a gang-related shooting. If that wasn't enough to clutter his mind, he also was displeased with a $650,000 base salary that ranked 18th on the 53-man roster.

"Junior tried very hard last year, but he was dealing with more than most of us could handle," coach Bobby Ross told the *San Diego Union.*

His biggest concern was his wife and daughter. One Sunday morning after Junior and Gina returned from church, her water broke, two months before her delivery date. They went to the hospital, where she was put on bed rest. The plan was to wait for Sydney's lungs to develop, then induce delivery. Even if everything went as planned, Sydney would need to spend a month in the neonatal intensive care unit.

Junior, who was participating in training camp, tried to focus on football, but his mind clearly was elsewhere. "He came and visited me every day, because I couldn't go home," Gina said. "Several times he'd sneak the dogs in the back door. He'd bring dinner to me, and we'd eat together in the hospital."

In 1994 his mind was free of distractions. Gina was fine, Sydney was fine, and his wallet was fatter after signing a $16.3 million deal in February that made him the league's highest-paid linebacker and its second-highest-paid defender, behind Packers end Reggie White. His play early in the year made it seem like the Chargers had gotten a bargain. He was excelling thanks in part to personnel upgrades around him, like new starting linebackers Dennis Gibson and David Griggs.

That they were added wasn't shocking considering the Chargers didn't get a single sack from that position the previous year; overall, San Diego had ranked last in the league, with only 32 sacks, in 1993. The defensive line also was improved. Leslie O'Neal, whose 29 sacks over the previous two seasons were third only to Kansas City's Neil Smith and Denver's Simon Fletcher, was already one of the game's top ends/ outside linebackers, and moving Chris Mims from tackle to end gave San Diego a formidable pass-rush tandem. In addition, putting tough-to-move tackle Shawn Lee next to Reuben Davis on the interior made it tough for opponents to run the football — and for blockers to get to Junior.

Still, there was no indication in the preseason that the Chargers were headed for a fast start. They not only lost their first four games but also got dressed down by Ross after absorbing a 31–3 spanking from the Oilers early in that skid. "Exhibition season or not, that was a damn lackluster performance," Ross told the media. "It wasn't only lackluster, it was poor in how we played, and I'm embarrassed. We can't turn the ball over five times, which led to scores. We can't drop seven to eight passes, and our protection was shoddy."

The preseason games were not unlike what Ross saw in the opening quarter against the Broncos in week 1. The Chargers could do nothing right, but they kept grinding until momentum turned in their direction.

At the halfway point of the season they were 7-1 and cruising when turbulence hit in the form of injuries. Humphries, who had missed the final series in a 20–15 loss to the Broncos two weeks earlier because of a badly sprained ankle, sat out a 10–9 loss at Atlanta. Center Courtney Hall was also dealing with a partially torn biceps and bum knee that would require surgery at the end of the season; and guard Joe Cocozzo had a foot injury that caused him to miss time. Even Junior was having issues. On November 20, on the fourth play of a 23–17 loss at New England, he sustained a pinched nerve in his neck. Each time he sustained a hit in that area, a current of incredible pain would shoot down his spine and through his body, leaving his right arm limp and useless at times.

"You've got to play through it," he told reporters two weeks after sustaining the injury. "You never play this game 100 percent healthy. It's nothing to complain about. I'll still be out there."

He didn't miss a game that season and helped the Chargers advance to the AFC Championship Game for only the third time in franchise history, but late that season it was clear he was not in a good place physically. Keoki Kamau, the Chargers' head trainer, was so concerned that he took Junior's helmet in one game and told equipment manager Sid Brooks to lock it away.

"I can't believe you'd just take my helmet," Junior said to Kamau, furious.

"Bug, at some point we've got to stop this," Kamau said. "I'm tired of seeing you hurting. I don't like seeing you hurting this thing even more."

"I'll be fine," Junior said defiantly.

Players often talk of a "warrior mentality" and never giving in to pain. They'll take shots, oral medication, anything to stay on the field, particularly when they see a player of Junior's stature sacrificing his body.

"There's a ripple effect," Kamau said. "John Parrella, the stud defensive tackle, breaks his thumb in the conference final and I tell him, 'You can't go back in.' He's like, 'Bullshit! I'm playing. Let's go. All we have to do is put a cast on it.' He went back in, and the next day he had surgery and they put pins in it. . . .

"When you look at what goes on, you constantly ask yourself, are you doing the right thing? Am I putting them in positions where they're going to get exposed to further injury? So you do all types of testing to make sure that that doesn't happen. Medically, anatomically, whatever test is available, doctor's experience — you use all you can to make sure they're okay."

The Chargers' coaching staff knew that Junior wasn't 100 percent for the showdown against the Steelers in the AFC Championship Game, so they designed a game plan that would accentuate what he still could do effectively. If they were going to send him on the blitz, for instance, they would have him come from an angle that would lessen the chances of him aggravating the pinched nerve. But like some plans, things didn't go as designed.

Early in the first quarter Junior walloped running back Barry Foster. An electricity-like current immediately ran through his body, leaving his right arm limp. Feeling eventually returned to it, but the sequence would repeat itself multiple times over the course of the game; each time it did, Junior refused to give in. He finished the first half with nine of his game-high 16 tackles as the Chargers went on to upset the Steelers, 17–13, and reach their first Super Bowl.

None of Junior's tackles was bigger than the last one: on third-and-goal from the 10 with 1:22 to play, quarterback Neil O'Donnell looked for fullback John Williams in the flat. The 231-pound Williams had caught a similar pass earlier in the game and bulled through linebacker Dennis Gibson for a 16-yard score. This time, after catching the ball at the 5, he had to deal with Junior — and Junior's bad arm. It was no contest. With the help of Griggs, Junior brought him down at the 3-yard line. One play later Gibson batted down a pass at the goal line to preserve the victory.

For Junior, succumbing to injury was not an option. Too much was at stake. The Chargers had been to the AFC Championship Game twice before but lost both times, in 1980 and 1981. The first defeat was 34–27 at home to the Raiders, whom they had split with in the regular season. The second was 27–7 to the Bengals, in what came to be known as The Freezer Bowl, a game that featured a wind chill of minus-59, making it the coldest game in league history. If that hadn't been disadvantage enough for the California-based Chargers that year, they had played the previous week in the high heat and humidity of Miami, prevailing 41–38 in overtime. In one week they had gone from the steam room to the icebox, from 79 degrees at kickoff to minus-9.

"It's pain," he said of his shoulder after the victory, "but after what happened here, it's worthwhile."

When Gibson batted down the goal-line pass, Junior didn't know whether to smile, cry, or laugh with glee. He did all three. Gina was the only immediate family member to attend the game, and when she reached him outside the locker room they hugged and sobbed. Her face was smashed against his jersey, which was drenched with sweat and rain, and all he kept saying was, "We did it! We did it, G!"

Gina had to cut short their time together because she needed to catch a plane back to San Diego. Junior reached town before her and, along with the rest of the team, immediately was taken to San Diego Jack Murphy Stadium, where more than 70,000 fans were waiting to continue the nightlong celebration. He was flanked on each side by his parents as he emerged from the tunnel to a large ovation. He wore a Chargers jacket, an AFC Championship Game cap, floral shorts, and Flojos. He said nothing, but it was apparent that he was practicing the teachings of his mother, who after each victory would tell him: "Put it in your heart. Don't show off. God has given this to you. Put it in your heart."

His father smiled and silently thanked God. "Half for my son, half for the team," he said.

All the negativity of Junior's rookie season was gone, and so was the losing that had preceded his arrival and lingered through his first two seasons. He had the respect of the locker room, his counterparts on other teams, and the community. He was much more than a football player. He was a brand, having started an active-wear line called Say Ow! Gear in 1993. At media day during Super Bowl week, at least 15 San Francisco 49ers wore some form of Say Ow! Gear, including quarterback Steve Young and former teammate Gary Plummer. It was an incredible sign of respect from an opponent, and the fact that it occurred on the biggest stage in professional sports made it even more special.

That same week Junior was named the NFL's "Man of the Year" for his work with at-risk youth, in addition to being the official spokesman of Pop Warner Football and the NFL kids' program. His star could not have been brighter. He even had national marketing deals with Nike and Frito-Lay, accounts that traditionally had been reserved for offensive players. But Junior's playmaking and personality transcended manufactured boundaries.

"Junior has charisma," said Steve Rosner, an executive vice president of Integrated Sports International who was working with Seau at the time. "He's emotional on the field, but he's not outrageous. People like that, and he's respected by his peers."

"It's warming to me to know that a beach bum like me can do things like this," Junior said. "It's always been a dream. [But] don't let all this fool you. It's great to have the media around and the commercials and endorsements and everything. But it's just a window within my life span . . . I just hope after I hang up my helmet, I'll be respected with the elite group."

The dream of playing in the Super Bowl turned out to be a nightmare. The Chargers were destroyed, 49–26, with Young throwing for a record six touchdowns. The outcome wasn't necessarily a surprise; in order to get over the hump after losing in the NFC Championship Game the previous two years, the 49ers had spent the off-season signing cornerback Deion Sanders, defensive end Richard Dent, and linebacker Rickey Jackson—free agents who all would later be inducted into the Pro Football Hall of Fame. San Francisco finished a league-best 13-3 in the regular season and was an 18.5-point favorite entering the Super Bowl, the largest spread in the game's history.

The Chargers had no answer for them. On the third play from scrimmage, Young found Jerry Rice on a post route for a 44-yard touchdown. On the fourth play of their next possession, Young located running back Ricky Watters, who converted a short pass into a 51-yard touchdown. And just to show they were more than big-play artists, the 49ers put together a 10-play drive for a 21-0 lead on their next possession.

Afterward, through the confetti that fell over Miami's Joe Robbie Field, Junior peered through his pain and thought he saw better days ahead for the Chargers. In fact, his future would be filled with repeated disappointments, as the Chargers would not win another playoff game during the remainder of his time in San Diego. In fact, they reached the postseason only once over his final eight years with the franchise.

9

Cheating . . . in Business and Matrimony

JUNE 29, 1996.

It was not your typical summer evening in San Diego. Instead of being mild, the temperature climbed from comfortable to hot to sweltering. It was one of those times when homeowners along the coast kicked themselves for not having air conditioning.

Things were even hotter on the northeast corner of the Westfield Shopping Mall, in an area known as Mission Valley. The prelaunch for a new sports-themed restaurant was taking place, and the "MVP Kickoff Party" featured some of the area's most popular professional sporting figures, including the man whose surname was on the front of the two-story, 14,000-square-foot establishment: Junior Seau.

It was an exhilarating and humbling experience for a kid who grew up poor in Oceanside, with parents who worked multiple jobs to make ends meet. In the nine years since he had graduated from high school, he had gone from owning little more than his good name to owning the city. Everyone wanted a piece of him, to be associated with him, to be near him. He was the Michael Jordan of San Diego, which made owning a restaurant fitting considering that Jordan, the iconic Bulls star, had done the same three years earlier in Chicago. The sports-bar market in San Diego was relatively uncrowded in 1996, and the idea that customers might get to see — or even meet — the celebrity owner figured to be a major draw.

Junior and his partner, John W. Gillette Jr., had architects model the place after sporting venues such as Denver's Coors Field, Cleveland's Jacobs Field, and the Chargers' San Diego Jack Murphy Stadium. The patio area had replica bleacher seats, and a 25-foot-long tunnel-like entrance simulated the long passage from the Chargers' locker room to the playing field. A long green mat with painted hash marks replaced the red carpet to further embellish the sporting theme.

If you hadn't known better, you'd have sworn it was a Super Bowl party. There were giant laser lights, stretch limos, screaming fans, and representatives from every media outlet. NFL luminaries such as Greg Lloyd, Duval Love, and Darren Woodson attended, as did Hollywood starlet Lela Rochon and San Diego mayor Susan Golding. As the party roared into its later hours, the lights dimmed and Alan Parsons Project's "Sirius"—the same song played for Chicago Bulls player introductions—blared through the sound system before a curtain fell to reveal a 25-foot-high mural of Junior, in full uniform, crouching and moving forward, as if to make a tackle. Lightning crackled behind him, making the image even more striking. If the goal was to affirm his larger-than-life status, he had succeeded.

The next morning the local newspaper described the place as big league, inside and out. There was sporting memorabilia everywhere and TVs at every angle, including a 150-square-foot projection screen that was the centerpiece of the restaurant. The place was everything Junior had hoped for after his previous big-picture business ventures failed or flamed out. "Say Ow! Gear" all but disappeared two years after its launch when the company was sued by a New Mexico sportswear production company that accused it of running off with a clothing franchise deal, according to *U-T San Diego*. In 1995 he debuted "Say Ow!" water in one-liter and half-liter sports bottles, but the venture quickly dried up.

Seau's The Restaurant would be different, Junior told himself, largely because of what he was hearing from Gillette, a financial adviser known for connecting with clients on both a business and a spiritual level, even going so far as to pray with them in his office on occasion. "I'm a Christian, [and] John looked me in the eye and said he lived by Christian values," said Ryan Jaroncyk, who had received an $850,000 bonus as a 1995 first-round pick of the New York Mets and turned over $100,000 to Gillette. "He had a picture of Jesus as a judge on the wall."

In reality Gillette was a phony who conned a number of high-profile athletes out of $11 million before pleading guilty in 1997 to 38 counts of grand theft and forgery. "Somewhere along the line you lost your moral compass," San Diego Superior Court judge Bernard Revak told Gillette before handing down a 10-year sentence, of which Gillette served just four years behind bars.

A Stanford graduate, Gillette worked in the financial business for many years, including stints as a vice president at Bank of America and a stockbroker at Shearson Lehman. At the latter job, one of his employees in the summer of 1991 was Patrick Rowe, then a star receiver at San Diego State. In a somewhat foreboding quote, Rowe told the *San Diego Evening Tribune:* "I don't know where this leads for me, but I do know if I apply the things that John teaches me I'll never be a person who has money and then turns away and loses it all."

Gillette set up his own operation in 1993, and two years later he and Junior decided to go 50-50 on a sports-themed restaurant near the stadium where the star linebacker wowed fans. Although Junior was the biggest name in Gillette's pond of professional athletes, he wasn't the only one. Chargers safety Stanley Richard, San Diego Padres pitcher Greg Harris, and triathlon champion Mark Allen were among his other clients who allowed him to handle large sums of their money. Allen and Harris were among the first to file suit against Gillette, claiming that money they had entrusted to him was unaccounted for. Other athletes immediately began to worry about their investments, and a short time later Junior was among those filing to have Gillette's operation placed in Chapter 7 liquidation bankruptcy. Before the end of August 1997 — just 14 months after the glitzy restaurant opening — Gillette was behind bars.

Gillette told most of his clients that their money was going into real estate investments, and in some cases that was true. But even when the investments made money — a Ghana gold mine investment was one that did not — most of the clients never saw the profit. Gillette kept those to himself to finance his lavish lifestyle, which included a five-bedroom home on a golf course and a Mercedes roadster. Though much of the money was unaccounted for, slightly more than $2 million was used for Gillette's share in Seau's The Restaurant. As part of restitution, Gillette surrendered his stake in the restaurant back to Junior.

On the field Junior continued to perform at a high level and helped the Chargers to a 4-1 start. One of the more memorable games of his career occurred that September, when the Chargers hosted the Kansas City Chiefs and future Hall of Fame running back Marcus Allen. Junior and Allen were close friends. Both grew up in San Diego County, both were All-Americas at USC, and both were iconic NFL players. Playing against Allen was special for Junior because he had so much respect for him. But this encounter had even more significance because the Chiefs were 4-0 and the Chargers were 3-1. First place in the division was on the line.

Junior put on a show in the 22–19 victory. He had 12 tackles, two interceptions, and a late-game sack on which he bulled through Allen, one of the game's best pass-blocking running backs. It was a stunning play because no one ragdolled Allen. He was too technically sound, too tough, too fearless. In college Allen actually played fullback before moving to tailback. But Junior treated Allen as if he were a blocking dummy, slamming into him before tossing him aside en route to the quarterback.

When he rose and pumped his fist into the air, the sellout crowd stood and roared with him. It was the highest of highs for Junior, winning the battle against an undefeated opponent and a future Hall of Famer for whom he had tremendous respect. But trouble loomed.

Behind closed doors in the Chargers' offices, general manager Beathard and head coach Bobby Ross were beginning to butt heads over personnel. The 4-1 start was followed by three straight losses, all within the division. Tension was rising inside the building and among the fan base. Two straight wins followed, but the Chargers lost four of their next five and finished 8-8. Shortly afterwards, the only coach to have taken the franchise to a Super Bowl was forced to resign.

Publicly, Beathard still had the respect of the locker room and community. Privately, people wondered about the decision to force out the coach who had brought them out of the darkness of nine straight nonplayoff seasons by taking them to three postseasons in his five seasons. The only way to ease tensions was to hire a capable replacement and win immediately.

On January 18, 1997, the Chargers announced the hiring of Kevin

Gilbride as their new coach. Gilbride, the Jacksonville Jaguars' offensive coordinator, had no NFL head-coaching experience but was known for being creative and innovative with his game plans — something that could never be said of San Diego's ground-oriented offenses under Ross. Hopes for a fast start were buried, however, beneath the rubble of a 41–7 loss at New England in the season opener.

Junior didn't play that afternoon because he was rehabbing from knee surgery. The injury occurred during the preseason, when Junior was run into by fellow linebacker Kurt Gouveia. He thought so little of it at the time that he didn't bother checking on the test results after undergoing an MRI. He had tested the knee a day earlier for peace of mind and came away thinking it was nothing more than a bruise.

"Everything is fine," he told Gina.

Even when he saw the blinking light on his answering machine he didn't think anything different. People were calling, it turned out, to say they'd heard he needed surgery to repair a damaged meniscus in his knee. TV reports confirmed it. The prognosis was that he could miss as many as six weeks. Junior was stunned. He had missed just two games in his first seven seasons. He was so down that he failed to return any of the calls from family and friends offering support.

"Even my mother couldn't reach me," he said. "I definitely had to come to a reality that I'm going to have to sit and wait this out. To have my wife and people wait on me — I'm not used to it. I'm a guy that loves to run and loves to do a lot of things — have fun, have a good time. But right now there's a setback in life, and dealing with it is probably going to be the hardest thing for me. It has nothing to do with going out there and playing. It has to do with just walking. It kind of humbles you. It really does."

Could he have made a difference in the opener? Probably not, but still, it was clear the Chargers missed their captain. With a week 2 game at New Orleans staring them in the face, there was a sense of dread within the organization about a 0-2 start if Junior didn't play. The game would mark Mike Ditka's home debut as coach of the Saints, and everyone knew the Superdome would be amped with excitement.

No one pressured Junior to play — he was less than two weeks removed from surgery, after all—but it was understood that his presence was needed. He was the life preserver who could keep them from going under. But would he play? Even as the players went through warm-ups for the game, no one was sure. Typical of Junior, he not only played but finished with 11 tackles, one sack, a forced fumble, and a fumble recovery in the 20–6 victory.

"Junior meant a world of difference," said veteran newcomer William Fuller. "Like I've told you guys before, I looked forward to playing with him. He's going to give it his all. It's those instances when you're tired, and you don't know if you can go, you turn around and you see him, you see the fire and the desire in him, and it'll pick you up and let you dig deeper and go a little bit further than you thought you could go. He's just a fierce competitor. Even though he wasn't quite 100 percent, at 90 percent he was still going to give you 110 percent."

"This guy was wired to the point that he probably had more nerves than the normal person," Kamau said. "I call these guys neurological geniuses because they could overcome so much. It's something that you and I can't understand."

His impact began even before the first play from scrimmage. He gathered the players in the locker room and spoke to them about what awaited them and what they needed to do to prevail. The players had heard such a speech countless times before in college, high school, and Pop Warner. The one that focuses on teamwork and playing hard and players are asked to give more even when they feel they have nothing left to give.

"He talked about a lot of the normal stuff," Fuller said. "But when Junior says it, you can just see the passion, the love for the game."

Predictably, Junior refused to accept credit for doing anything out of the ordinary. "There are pieces to a puzzle, and I'm a big piece to this puzzle," he said. "Because of that, I'm expected to do the things I did out there whether I'm healthy or not. It worked to our advantage and for me personally. But for me to be the man that revived this team—no, I'm not going to take credit for that."

Privately, however, Junior acknowledged the tremendous internal pressure he felt to play . . . and play well.

"The public are our employers, but they only know a person named Junior that is a winner on the football field," he wrote afterward in a journal entry to Gina.

> They love the intensity that we play with on every Sunday afternoon; they also love the person because of the perception of a family man. To earn all this is a big accomplishment, but when you need understanding they don't have time. I leaned on you with all my heart and soul to help guide me back with a positive [result]. To be injured and play average, you are failing in everyone's eyes. Why? Because we convinced everyone that I'm a player that should win. This week was a do-or-die situation for me.

Unfortunately for the Chargers, the victory did not jump-start them. They lost their next two games to fall to 1-3. The frustration was palpable within the training complex. Even Junior, who refused to publicly acknowledge negativity, showed signs of cracking—as he did when asked about the defensive scheme under new coordinator Joe Pascale.

Because of injuries within the linebacking corps, Junior was spending most of his time at middle linebacker, where he was being caught in the congestion that came from dealing with centers and guards. Why not move him on the edge to give him some freedom?

"I'm just following the scheme," he said, which was as close to a public second-guessing as you could get from him. "The scheme is to go inside, and that's what I do."

Is that what he preferred?

"I just follow the scheme," he said. "I come from wherever they want me to come from."

But he acknowledged that more losses would test his patience. "It's all going to come to a head," he said.

The reality was that Junior was frustrated with more than his role on the field. He also was bothered that attempts to reach agreement on a contract extension were going slower than he wanted. His surgery a month earlier had reminded him how quickly a career could end, so he wanted to cash in at least one more time to protect himself.

Junior was smart when it came to the business of football. He knew he had leverage. The Chargers had just fired a popular and successful coach only to replace him with someone who was now 1-3. The fan base was uneasy, and the last thing the organization wanted was the face of the franchise speaking out against it. So he said just enough to get management's attention.

Less than two months later, with the franchise headed toward a 4-12 finish that would match its worst record in six years, Junior signed a $26.1 million extension that included a $6 million signing bonus. The deal made him the league's highest-paid linebacker with an average salary of $4.5 million a year, and it allayed concerns within the organization that its most visible and talented player might be a problem in trying to turn things around.

Although the contract brought peace of mind from a financial standpoint, Junior was struggling in his personal life. He was feeling more tension between his wife and his family members in Oceanside, some of whom felt that Gina was guilty of taking him from them. He had become a human rope in an escalating tug-of-war, and it was becoming increasingly difficult for him to persuade family members that Gina loved him for the person he was rather than the fame and fortune he represented.

"We let the closet stay closed because of Junior, but she made it very difficult," said Annette, his younger sister. "When the two of them first met, they used to come visit my parents. But the minute he put a ring on her finger, we hardly saw Junior. The thing that really bugged me the whole time was the fact that she would not communicate with anybody in the family other than me — and the only reason she communicated with me was because of Junior. My husband and I didn't have kids at the time, so we spent time together with him and Gina. I was cordial toward her because of my brother, but I just didn't like how she was treating my parents and my brothers and sisters. Junior felt a lot of strain. He tried so hard to have the family barbecues at his La Jolla house, but she wasn't cordial with the family. She only stayed with her people that she invited. You were able to see it. They would get into arguments because we would stay late."

Gina was unlike anyone he had ever dated. And while it would be nonsensical to say that he didn't love her, it likewise would be foolish to deny that part of the attraction was his belief that she could help him navigate a social world he was completely unfamiliar with—which in turn might erase the perception that he was a "dumb jock" who couldn't secure a qualifying score on a college entrance exam.

There appeared to be bliss in the early years of the marriage, but the tension between the families was always present. It escalated after Sydney and Jake were born, when some family members believed that Gina was keeping the kids away from Oceanside because she wasn't comfortable with the area or with them. The only way for the grandparents to spend time with the kids, they said, was for Junior to bring them to Oceanside or for Mom and Dad to visit his home. Complicating matters was that Junior and Gina were having problems of their own. She was becoming increasingly suspicious that he was cheating on her.

Junior had gone from being a husband who wrote spiritual affirmations of his love on Post-its for Gina to find when she got out of bed to being a husband who was distant and evasive. It became most noticeable after the couple moved from the inland Mount Helix area to the exclusive coastal community of La Jolla. Junior joined La Jolla Country Club and started hanging out at The 19th Hole, the club bar. He was harder to reach by phone and dismissive when home. When he returned from road trips, there were phone numbers from unknown women in his suitcase or his car, as well as handwritten notes and phone numbers from women at his restaurant.

Junior may not have been searching for trouble, but he was about to find it.

Junior had five brothers and sisters. This picture, taken in the early 1970s, shows all of them before the birth of Antonio. Back row: Annette, Savaii, Mary, and David. Front row: Junior.

Bette Hoffman

Junior had to sit out his first season at USC because he failed to achieve a qualifying score on the entrance exam. He was a reserve most of his sophomore season (his first on the field), then became an All-America in his third season, with 19 sacks and 27 tackles for loss.

Bette Hoffman

In 1990, the Chargers selected Junior fifth overall in the NFL draft, fulfilling his boyhood dream of playing for the franchise he grew up supporting. At his introductory press conference with the San Diego Chargers—many of whom he already knew—he was flanked by coach Dan Henning (Junior's right) and general manager Bobby Beathard (Junior's left).

San Diego Chargers

Junior sat out nearly all of training camp as a rookie because of a contentious contract negotiation. When he did report, it was the week of the final preseason game. He entered late in the first quarter but was ejected after two plays for punching Raiders guard Steve Wisniewski. Here he walks to the sideline after being kicked out of the game. *San Diego Chargers*

Two years after meeting, Junior and the former Gina DeBoer married in March 1992. The couple had three children before divorcing after a decade of marriage. *Gina Seau*

Junior consistently led the Chargers in tackles during his 13 seasons. As this image of him striking a Bengals running back shows, he had to be accounted for at all times.

San Diego Chargers

Junior Seau holds up the Lamar Hunt Trophy after the New England Patriots defeated the San Diego Chargers 21–12 to win the AFC Championship at Gillette Stadium on January 20, 2008. The win gave the Patriots the first 18-0 record in NFL history and sent them to the fourth Super Bowl in seven years. *© David Bergman / Corbis*

During a 2001 practice at training camp, Junior collided so forcefully with fullback Fred McCrary that it caused a three-inch crack near the crown of McCrary's helmet. Later that night Junior told McCrary, "My head is on fire!" *Jim Trotter*

When the Chargers decided to part ways with Junior, the Dolphins were high on his list because their roster was dotted with Pro Bowlers, which meant he'd finally have another shot at the postseason. It also didn't hurt that the Dolphins played on grass and in warm weather. *Miami Dolphins*

Junior had three children with Gina Seau. From left to right: his son Hunter, daughter Sydney, and son Jake. *Bette Hoffman*

Believing that his career was over, Junior participated in a 2006 retirement ceremony hosted by the Chargers. He refused to say he was retiring, however, telling the large gathering that he was graduating. Four days later he signed with the New England Patriots.

Mike Norris Photographer

Junior Seau hugs San Diego Chargers teammate Rodney Harrison after the game against the Kansas City Chiefs at the Qualcomm Stadium in San Diego, California. The Chargers defeated the Chiefs 17–16.

Stephen Dunn / Getty Images

Junior Seau talks with head coach Bill Belichick during New England's 34–17 win over the Cleveland Browns on October 7, 2007.

AP Photo / Winslow Townson

Before playing the Chargers in San Diego in October 2008, Patriots owner Robert Kraft and several team members visited Junior at Seau's The Restaurant, where Kraft presented him with an AFC Championship ring from the previous season. Junior was unsigned at the time of the visit, but rejoined the Patriots later that season.

Bette Hoffman

Junior Seau walks off the field as confetti falls after the New York Giants' 17–14 win in the Super Bowl XLII football game at University of Phoenix Stadium on Sunday, February 3, 2008, in Glendale, Arizona. *AP Photo / Paul Sancya*

Junior was the emotional heartbeat of every team he played on. Most of his coaches allowed him to address the team before taking the field for games. Here, he imparts words of wisdom to Chargers players—some of whom he played with—before participating in his retirement ceremony.

Mike Norris Photographer

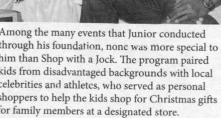

Among the many events that Junior conducted through his foundation, none was more special to him than Shop with a Jock. The program paired kids from disadvantaged backgrounds with local celebrities and athletes, who served as personal shoppers to help the kids shop for Christmas gifts for family members at a designated store.

Mike Norris Photographer

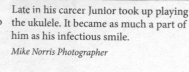

Late in his career Junior took up playing the ukulele. It became as much a part of him as his infectious smile.

Mike Norris Photographer

Junior's car is hauled onto a truck after it was driven off an embankment October 18, 2010, in Carlsbad, California. Seau was hospitalized with minor injuries.

AP Photo / KFMB-TV

Tire marks are clearly visible on the cliff where Junior Seau drove off an embankment.

AP Photo / Gregory Bull

During halftime of a November 2011 game against the Broncos, the Chargers inducted Junior into their Hall of Fame. From left to right: Mama Seau, Junior's son Tyler, his daughter Sydney, his son Hunter, club chairman Dean Spanos, and Papa Seau.

San Diego Chargers

Junior always had a moment of worship at the banquet preceding his annual golf tournament. But in 2012, two months before his death, he did something he had never done: He called his family onto the stage to worship with him. *Mike Norris Photographer*

Junior spent many of his mornings in the Pacific Ocean, which was a few steps from his beachfront home in Oceanside, California.

Mike Norris Photographer

Papa Seau is comforted by his daughter Annette at the ceremony to retire Junior's jersey following his death. *Mike Norris Photographer*

On the one-year anniversary of Junior's death, family members gathered at the gravesite for what's known in Polynesian culture as the crossover. *Jim Trotter*

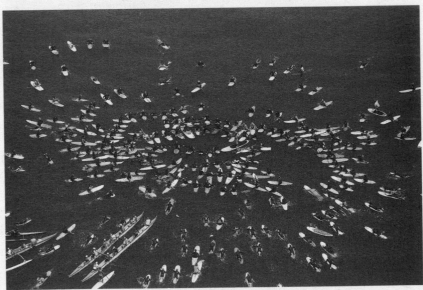

Hundreds of fellow surfers remembered Junior Seau during a traditional paddle-out ceremony in front of his Oceanside, California, home on May 6, 2012.

Read Miller/Sports Illustrated/Getty Images

10

The Anti-Leaf

HIS NAME was Ryan Leaf, but in San Diego he was viewed as The Savior.

Desperate for a quarterback after their four signal-callers combined to average less than one touchdown pass per week during the 1997 season, the Chargers traded up one spot in the 1998 draft and selected Leaf, a six-foot-five Washington State gunslinger with a big right arm and bigger mouth, with the second pick overall — one spot behind Tennessee quarterback Peyton Manning.

NFL personnel people were split going into the draft as to which would be the better pro. Manning had the pedigree and the polish. His father, Archie, spent 14 years in the NFL after being drafted second overall by the Saints in 1971 and was considered football royalty. Peyton, who grew up around the game, possessed many of his father's qualities: he was hardworking, competitive, and mature beyond his years. He even showed up for predraft interviews wearing a tie and blazer and carrying a notebook so he could record answers to the questions he had brought for his potential employer.

Leaf was the opposite. He grew up in Great Falls, Montana, and never got closer to the NFL than his television screen. He was known for being loudmouthed and boorish, so much so that some of his high school and college teammates admitted they weren't overly fond of him. That statement was jarring considering Leaf played the most important position on the field and often was the best player on either sideline. But he got away with it because he delivered the football regardless of rain, sleet, snow, or opponent.

The Chargers didn't particularly care which quarterback they came away with; they just had to ensure they'd get one of them. Beathard did that by trading two first-round picks, a second-round pick, and three-time Pro Bowler Eric Metcalf to Arizona for the right to move up one spot in the draft. After the Indianapolis Colts selected Manning, the Chargers eagerly claimed Leaf, eliciting cheers and talk of brighter days among fans.

Junior was ecstatic with the selection. He knew the chances of winning a championship increased significantly with a franchise quarterback, and as much as he respected guys like Stan Humphries, Jim Harbaugh, Erik Kramer, Jim Everett, Craig Whelihan, Todd Philcox, Sean Salisbury, and Gale Gilbert—quarterbacks who had taken snaps for the team over the previous three seasons—Leaf had the arm, bravado, youth, and college résumé to be a difference-maker.

But Junior had been the dominant face of the franchise for nearly all of his career. People immediately wondered: how would he take to a teammate stealing some of the shine from his star? The question went from a whisper to talk-radio shouting two months later when Junior skipped the start of a voluntary team minicamp, something he had never done before. In fact, the move had more to do with money than jealousy.

The contract that Junior had signed eight months earlier had essentially become obsolete because of a sharp increase in the salary cap, which created space for bigger contracts than the one he had signed. Junior wondered if the team knew the cap was going to jump and purposely signed him before it did so.

"Junior's got something that he feels strongly about, and he's gonna deal with that in his own way," said middle linebacker Kurt Gouveia. "If you felt strongly about a situation or something that bothered you, and the only way to express that is by doing what he's doing, I don't blame you. Obviously he's one of our leaders and we want him to be here. But sometimes you have to step back and say, 'Man, he really feels strong about something that he's upset with.'"

Gouveia may have been understanding, but Junior's absence seemed to catch everyone else off guard. Gilbride said he was "shocked" that the team's leader wasn't present, and Beathard called the move "selfish."

"One of the great qualities of leadership is not being selfish," Beathard said. "Football is a team game, and it takes every single part to keep making progress to get to where we want to go. These kinds of things slow down progress and set things back. We're not at a stage where we can afford to do that."

The characterization that he was selfish stung Junior deeply. He had always considered himself a team player. He was there for anyone who needed him, and he never spoke out against the organization at times when nearly everyone outside the franchise was doing so. Now his character was being attacked by the same people he felt he had protected?

He allowed the Chargers to control the narrative for a full day, then hired a PR firm to perform damage control. The agency issued a statement from Junior in which he reminded everyone that the minicamp had been voluntary. He also stated: "Due to a rigorous travel schedule, I will be away from home for most of the month of June. This is my only opportunity to enjoy a short family vacation."

The fallout was so severe, however, that he returned to San Diego for the final two days of the camp. He claimed that he was fine with his contract, but it was clear during an uncomfortable media session that he had issues with his coach and general manager.

"I love Bobby," he told the media. "Bobby and I, we're boyfriend-girlfriend. That's my girlfriend, and I'll apologize to her and we'll get back together. It's quite all right. My head coach? That's my other girlfriend. They get kind of temperamental, you know? . . . We'll come to a point where we understand each other."

"I don't even understand that, much less [feel] able to answer it," Gilbride said when asked about the comments. "Sometimes when you don't have a lot to say, you say some things that don't make a lot of sense . . . When you're 4-12, you've got to be here. You've got to be willing to pay the price."

Privately, Junior seethed at the "pay the price" comment. What did that mean? For nearly a decade he had given the organization everything he had, physically and emotionally. He didn't rip management for forcing out the lone coach to take them to a Super Bowl so it could replace him with a purported offensive guru (Gilbride) who oversaw a passing game that produced a league-low 12 aerial touchdowns in his first season in San Diego. Pay the price? In Junior's mind he had done that 10 times over.

He wanted to lash out even more than he did in that awkward press conference, but he knew he couldn't. It would feed the perception that he was selfish and might cause the public to turn on him, as it did his rookie year when he held out. It also would have forced him to answer to his father. So Junior turned inward. He leaned more heavily on Gina and passed on nearly every interview request.

His silence didn't help matters, though, particularly after he laid a vicious block on Leaf during training camp. The rookie QB had thrown the first of what would be numerous interceptions in his career, and as the defense returned the ball down the field during the morning practice at UC San Diego, Junior intentionally leveled Leaf.

Onlookers gasped at the force of the hit. It's widely known that quarterbacks are not to be touched in practice. They typically wear red jerseys to remind teammates they're not to be hit. That awareness is raised tenfold when the quarterback has just received a franchise-record $11.25 million signing bonus and is considered a key to the season. What was Junior thinking? How could he risk injuring Leaf?

In Junior's mind it wasn't about jealousy; it was about respect and protocol. There's a code among players that issues between teammates are handled within the locker room, with management and coaches never being brought into the equation. Leaf infuriated teammates when he whined to management that a group of veterans had used his credit card to pay a $2,800 dinner tab and place a $1,500 bid in his name on a Qualcomm Stadium skybox.

In the players' minds, it was part of the rookie initiation process. Nearly every other high draft pick had been treated in a similar fashion during their first training camp. But by complaining to Beathard, Leaf had broken the code and needed to be reprimanded.

"When the guys found out Ryan had gone to Beathard, they were so pissed," said Billy Devaney, the team's director of player personnel at the time. "Junior wanted to send a message. He hunted Leaf down and de-cleated him. The whole defense came over and high-fived him right away."

Junior was old school—from the R&B tunes he often blared from a stereo system in the locker room to his respect for authority, to his attitude about the game. Issues involving players were meant to be handled by the players. When some veterans gave him the cold shoulder when he was a rookie, refusing to huddle for him after he failed to participate in training camp because of the contract dispute, he didn't run to Beathard. He kept working until he had their respect.

But now, for the first time since his rookie season, Junior found his character being attacked. Making matters worse was that he was dealing with a significant drama in his personal life. Gina had noticed irregularities with the phone records earlier that spring. She tried to ignore them, but when she began finding handwritten messages for him at the restaurant—from women she didn't know—her gut told her something wasn't right. She bluntly questioned him about it.

"A woman always knows if something's not right," Gina said. "You know the patterns and behavior. I confronted him. I told him, 'I don't know the extent of how much you're cheating, but if you want to continue this behavior, that's fine. You will not have me at your side to do it, though.' I was very direct. I said, 'I love you, but I think you're making poor choices. I'm not going to stand for it.' I threw all my cards on the table, everything that I had. I asked him, 'What's going on? Do you want to be with us, or do you want to be a bachelor? Just tell me.'

"He felt terrible and owned up to some stuff. A friend of his was caught cheating by his wife, and the next night I was going to visit his friend's wife, and he thought by me doing that I was going to learn everything he [Junior] was doing. The night before I went over there to visit her, I came home and he was waiting up for me at 11 or 12 that night. He was like a deer in headlights."

When he finally admitted to cheating, Gina's voice wasn't the only voice that sent him to a dark place. His father's did as well. He could hear the words from his childhood about the importance of honoring the family name. Junior felt ashamed that he hadn't been the man others believed him to be—and the man he wanted to be. He promised to change — to part company with destructive friends, spend more time at home, and attend counseling sessions. He also reconnected with the religious roots he'd established while growing up in the church.

The change in him was noticeable to everyone at the training facility. The meter on his larger-than-life personality had been turned down from 10 to 5, and his toothy smile had been replaced by a soft grin. He was respectful but distant toward the media, turning down nearly every interview request until Thanksgiving week, when he told a reporter that he was working hard to become a better teammate, husband, and father.

"I can't say he's a man of God now, as if he wasn't before," Gina said at the time. "I just think he's walking the walk better, not just talking the talk. He's a living testimony now of just being a good man, you know what I mean? There's a difference between saying it and doing it. Actions are so much more powerful than words, and his actions have stood up to his testimony of what he wants to do and what he wants to be."

Junior also directed more energy toward his foundation after hiring Bette Hoffman. The two met on a lark. Junior was having a luncheon to raise money for his foundation's "SACKS for San Diego" program, which asked donors to contribute up to $100 for each sack the Chargers recorded during the season. The money would fund programs to help kids avoid drugs and violence. Hoffman wasn't supposed to attend the event; she had had a previous meeting scheduled that afternoon. But when her appointment was canceled, the phone rang. A friend had a last-minute cancellation and wondered if she'd like to join her at Junior's luncheon. Already dressed for a business meeting, she said yes.

Hoffman didn't know Junior personally, but she was aware of him because he and Gina were members of the United Way Tocqueville Society, a charitable program that focuses on volunteerism as a vehicle to create life opportunities within at-risk communities. Hoffman had previously crossed paths with Junior while doing work for the United Way, but those encounters had usually consisted of nothing more than "Hello" or "How are you?" This time they had a substantive conversation after the event ended.

Junior asked what she thought of the luncheon. Her first thought was that it wasn't well organized. She was doing consultant work for nonprofits and told Junior she'd be happy to work on his fundraising committee, not realizing that there was no such committee. The next week Junior called and asked if they could meet. He wanted her to assist with his foundation.

Hoffman initially declined because she was busy with other projects, but Junior persisted. "'Nobody can say no to Junior,' he said. After about two or three meetings, I said, 'Okay, I'll give you one day a week.' Pretty soon I was overseeing the foundation and the restaurant and everything else that he had going on."

The foundation was operating far from peak efficiency at the time. Junior and Gina started it in a home office, and while their hearts were in the right place, neither had the business acumen to help it reach its full potential — despite Junior being named the NFL "Man of the Year" in 1994 for his community work. For instance, his big fundraiser each year was a weekend golf tournament that drew only 23 foursomes, half of what a tourney should attract at peak efficiency. Hoffman explained to him that the first order of business was moving the tournament from June to March; that way it wouldn't be on the tail end of the NFL fundraiser season. In two years the tourney doubled the number of participants and maxed out, at 56 foursomes.

Hoffman and Junior were a formidable team because one was strong where the other was weak. Junior had the star power and charisma to attract donors; Hoffman had the organizational and planning skills to stage quality events. However, it didn't take long for her to realize that Junior was a sponge when it came to business: he quickly absorbed all facets of fundraising. "I'd have a budget and projection formula that I work with for money coming in and how to manage it, and he got it immediately," she said. "I had worked with a lot of CEOs, and sometimes they'd have trouble understanding that part of it because they all had CFOs. Not Junior."

Typical of Junior, he wasn't satisfied with the foundation being good. He wanted it to be great, and in Hoffman he had someone who shared his vision for making it a game-changer and life-changer for kids and young adults. In no time the foundation was awarding 17 $1,000 scholarships, each renewable annually for four years. It also provided tutoring and mentoring for at-risk youth, including Pina Tinoisamoa, a talented but troubled athlete from the Oceanside area who was facing what amounted to a third strike after being locked up as a teenager.

The Junior Seau Foundation was sponsoring a "Gang Busters" program at the time, and two of its organizers spoke to the court on Tinoisamoa's behalf to secure his release. The program then worked with him to improve his grades. Ultimately he earned a football scholarship to Hawaii, and in 2003 the linebacker was drafted in the second round by the St. Louis Rams.

"The Gang Busters program helped set me straight," said Tinoisamoa. "With it being right in the community, it was something that saved my life. That was Junior. He knew how important something like that program was. He went through it with his brother Tony. He knew how prevalent gangs were, and he wanted to do what he could to give us alternatives, positive alternatives. He had people surround me and help set me straight so I could live my dream. They would pick me up from school and take me to their after-school program."

Over time the foundation would distribute some $4 million to various programs. It also would shut down Junior's restaurant each Thanksgiving to feed families of domestic violence and military personnel who couldn't be with their families. At Christmas it would hold "Shop with a Jock," one of Junior's favorite events. He'd bus kids from Oceanside to San Diego, where they'd receive $100 gift cards to purchase gifts for family members from the department store sponsoring the event.

The foundation helped distract him from the issues surrounding Leaf, whose poor work ethic and lack of accountability made him an outsider in his own locker room. Signs of trouble for the former Washington State star were there from the start. In addition to the negative scouting reporters, Leaf had persuaded owner Dean Spanos to fly him to Las Vegas the night he was drafted so he could attend a friend's birthday party. When he flew to San Diego the next morning for the introductory press conference, Leaf appeared hungover. Later, he skipped the final day of a mandatory rookie symposium and was fined $10,000 by the league office.

For Leaf, it was one misstep after another, the most memorable coming the morning after a week 2 loss at Kansas City. Upset that a reporter had written about his boorish behavior toward a cameraman following the defeat, Leaf erupted when the reporter spoke to him the next day. "Don't talk to me!" he yelled while standing over the reporter in a threatening manner.

At that moment, Junior could have allowed Leaf to dig an even deeper hole, but the perennial Pro Bowler silently stepped in and grabbed Leaf's arm before leading him away. He also didn't publicly attack Leaf six days later after Leaf threw four interceptions in the first three quarters and was benched in a home loss to the New York Giants. Instead, he said the team had to get better.

By then it was apparent that the season was not going to end well for Leaf or the Chargers. Leaf was in over his head, Gilbride was on the way to being fired just six games into the season, and the Chargers were headed for their third straight nonwinning year (they finished 5-11). Junior refused to let the team completely unravel, though. Despite being hindered by an offense that routinely compromised the defense by committing an astounding 51 turnovers — 22 above the league average — San Diego led all clubs in total defense and rushing defense. The tougher times got, the harder Junior worked.

"I once asked him why he practiced so hard," said Harrison, his teammate in San Diego and New England. "He said, 'Rodney, I get paid to practice. I play the game for free. Anybody can go out in front of 70,000 and get excited and play a game. But it takes a special person — not a special player, a special *person* — to practice at game speed.'"

When quarterbacks coach June Jones was promoted to interim coach following the firing of Gilbride, he sought to put his own stamp on the team. Jones was a morning guy, so he switched the practices from their customary afternoon slot to 8:00 AM. His expectation was that guys would show up at 7:30, get taped, get dressed, go to the morning meeting, hit the field, then do their weight training in the afternoon. However, when Jones arrived for his first full day on the new job at 4:30 AM, one player was already there. It was Junior.

Shortly after, Junior started what came to be known as the Breakfast Club. It was early-morning weight-training sessions that took place before players were required to report for work. There was no formal announcement. News of the sessions spread by word of mouth. If you were interested, you knew to be there by 6:15.

Junior was always the first to arrive, except one morning when Harrison got there at 6:10. That didn't sit well with Junior, who viewed it as someone outworking him. When Harrison showed up at 6:10 for the next session, Junior was waiting for him.

"Good afternoon, Rodney," he said.

Harrison was as competitive as Junior, so he arrived for the next workout at 6:00, five minutes ahead of Junior. Predictably, Junior showed up at 5:50 the next time, well ahead of Harrison.

"Finally I said to him, 'Junior, this has to stop before we're both getting there at four o'clock in the morning,'" Harrison said. "That's the type of guy he was. Always challenging people, always pushing people. With young guys, he would say, 'I can tell you everything you want to know, but rather than tell you I'm going to show you. Anything else is just lip service. They're just words coming out of my mouth.'"

As Thanksgiving approached, Junior led the club with 80 tackles but had just one and a half sacks, which tied for only sixth on the team. Instead of grousing, though, Junior nodded his approval because he had asked Joe Pascale, San Diego's defensive coordinator from 1997 to 2000, to use him as a decoy to help create opportunities for others. The request stunned Pascale, who knew Junior as a guy who not only wanted to make every play but *needed* to make every play.

When the two had first met in 1997, the team was moving from its longtime offices at San Diego Jack Murphy Stadium to a new training complex 10 minutes to the north. Pascale was unpacking boxes in a double-wide trailer serving as his temporary office when Junior walked in and skipped the perfunctory pleasantries.

"Tell me what we gotta do to be better," he said. "I'm ready. We can get these guys started."

Pascale was surprised. He and his staff were still moving desks and tables, with months to go until the start of off-season workouts. He had yet to break down the tape of the defensive personnel or to formulate a blueprint for how each guy would fit in the puzzle. But on this day the face of the franchise was standing in front of him, wanting answers — specific answers — for what they were going to do, how they were going to do it, and who they were going to do it with.

Pascale asked him to pump the figurative brakes. "That wasn't what he was looking to hear," Pascale said. "He wanted to know what he could do today to get this thing started in the right direction."

It wouldn't be the last time Junior pushed Pascale. Each Tuesday or Wednesday during the season he'd walk into Pascale's office and ask how Pascale planned to get him involved in that week's game. By that time Junior already had studied the next opponent's offensive line, running backs, and offensive formations and tendencies that he could exploit. If Pascale's plan didn't include things that would put him in the middle of the volcano, so to speak, he'd say: "What, are you going to have me watch the game?"

He was the same way on the sideline during games. Pascale liked to make certain calls early in games to see how the offense would react. Sometimes those calls, made consecutively, didn't put Junior in the middle of the volcano, and he'd grow impatient. He'd get the defensive call from Pascale between plays, and before passing it along to his teammates, he'd step away from the huddle and drop his hands to his side. Pascale knew what that meant: *When are you going to get me involved?*

"I knew if I didn't get him involved, he was going to start dancing around and get involved on his own," Pascale said. "Getting him involved was always part of our plan because he was a disruptor. That was the foundation of our defense — to disrupt the timing of the offense, make them feel uncomfortable, hinder their ability to make checks, and so forth. The biggest thing that impressed me about Junior overall is that he wanted to be in the middle of the fray when the game was on the line. Some people — I don't want to say they're afraid of it — but they're hesitant when the big moments come up. But he wanted to be in the middle of it, not just for the sake of the play, but because that's where he felt he belonged. That was the best part of the game for him."

The perception that Junior often played outside the defense had been valid during his early years, when he relied on his athleticism and instincts because he was still learning the league and opponents' tendencies. But as the years passed his "freelancing" had been built into the defense.

"A lot of being involved with him was blitzing and attacking that A gap," said Pascale. "It's very intimidating for a quarterback when Junior's in that A gap — one yard away — and the ball's going to be snapped, and the guard's got to block down on him. We knew the offense had to account for him and the protection was going to slide to him, so we used to run stunts off it. You couldn't put a back on Junior because the back couldn't get to him before he got to the quarterback. They had to slide their protection that way, so our other blitzes were based on that. We'd run the blitz away from the protection."

Pascale felt that he had at least one stud on every level with John Parrella at tackle, Junior at linebacker, and Rodney Harrison at safety, and he game-planned around them. For instance, he tried to always balance the field by having Harrison on the strong side and Junior on the weak side. He knew if he put them on the same side, offenses would have an easier time scheming away from them, and he was not going to allow that to happen.

Yet here was Junior, during his regular meetings with Pascale, now requesting to be used as a decoy. Here was Junior, the quiet amid the storm. But would it last?

Junior and other veteran players had grown close to Jones, the interim coach. Jones gave them a lot of input into how things were done, and he believed in treating players as if they were business partners instead of employees. One of the first things he did after replacing Gilbride was move the coaches and other nonplayer personnel to the back of the team plane for away games so the players could sit up front in the wider seats.

"It just showed the level of respect he had for players," Harrison said. "He always used to say, 'The players are the reason why we're here.' He showed an appreciation for us. It was a small gesture, but it meant a lot."

When Beathard asked Junior about keeping Jones as the full-time coach, Junior eagerly signed off on the idea. He not only liked Jones as a person but also appreciated the fact that he had Jones's ear. He was much more likely to get his way with someone as amenable as Jones versus a hard-ass like Gilbride. The only problem was that Jones wasn't interested in the job. His heart was set on the University of Hawaii, where his arrival was greeted with "Dewey Beats Truman" type headlines.

That left the Chargers searching for their fourth coach in as many years.

11

"My Head Is on Fire!"

THE YEAR 1998 had been a difficult one for Junior, both personally and professionally, so he eagerly looked forward to what the new year would bring. Nothing intrigued him more than what the Chargers would do about hiring a new head coach.

Beathard had had a track record of hitting the jackpot in such situations, like hiring future Hall of Famer Joe Gibbs in Washington or landing Bobby Ross out of Georgia Tech. But he had missed so badly on Gilbride that no one knew what to expect. When he chose Mike Riley from Oregon State, people scratched their heads.

Riley was an all-time nice guy, but he had no NFL experience and lacked a commanding presence. He spoke with a slow quasi-drawl and led "Hip Hip Hooray" chants in the locker room after victories. Junior was among those who had to do some research to learn about him; he quickly took to Riley because he saw qualities that reminded him of Jones, most notably that Riley would treat players with respect. It also wasn't a stretch that he saw an opportunity to heavily influence his coach's thinking. Riley admitted as much years later.

"I was ill prepared for that job," he said. "I had never coached in that league or even been a coordinator in that league, so [I lived] with the constant thought that it was unlike any job I'd ever had. I had been a longtime assistant in the Canadian Football League before I was a head coach there. Then I was an assistant coach at 'SC before I got the job at Oregon State, so I knew the league and I knew the teams. I felt confident. But in San Diego I constantly worried, 'Am I doing the right job for this team, the right thing in the NFL?' Then you add to that

that I'm getting the opportunity to be a coach for one of the greatest players to have played the game, and you wonder: 'Are we doing the right thing for him?' One of the things a coach has to do is turn around the perspective and look at it from a team standpoint. Look back at it as coaches and say, 'What do they need from us?' That was always a concern for me coaching that team and coaching Junior Seau."

Beathard selected Riley primarily for two reasons: he had strong people skills, and he had a creative offensive mind. It was hoped that Riley could use both those qualities to bring out the best in Leaf, whom the organization still hoped to salvage. Riley knew he would need the veterans to back him, so he spoke to Junior about the importance of trying to make a player out of the embattled quarterback. Junior agreed to do what he could to persuade the locker room to give him a second chance.

"He remained faithful to his team by trying to support Ryan Leaf longer than anyone did, talking to him, being supportive of him around the building," said Riley. "There were some rough times in there, but he knew the importance of the investment at that spot, that that was going to have to work for us to be any good. He stayed with it for a long time."

On most teams the first-string offense and the first-string defense don't practice against each other. Each works against what's known as a scout team, which is made up largely of bottom-roster guys who try to simulate the plays that that week's opponent might run. When the number 1 offense is on the field, the starting defense usually hangs out on the sideline waiting for its turn to take the field. Ditto the offense when the number 1 defense is on the field.

One day Junior excused himself and walked from the huddle to the sideline after seeing Leaf standing by himself. "Ryan was over there, and it was like the loneliness of the long-distance runner," Riley said. "Who knows what they talked about when Junior went over there? Maybe Junior had some personal empathy for the situation Ryan was in. But he also knew for the team's sake this thing had to work."

The attempt to develop Leaf turned out to be another brick in the road of good intentions going nowhere. Early in training camp Leaf began complaining about pain in his shoulder. The organization, with the backing of the players, decided not to waste time hoping he would come around. Instead, they turned to veteran quarterbacks Jim Harbaugh and Erik Kramer, a move that made Junior and other veterans ecstatic.

As much as he tried to support the organization, Junior knew all along that Leaf lacked the heart and commitment to excel. Harbaugh and Kramer were well past their prime, but at least the game mattered to them and they could be counted on to give their all. The reality that the team would not be weighed down by Leaf invigorated Junior. So did Riley's decision to allow Junior to take reps at tight end in training camp.

By that point, entering his 10th season, the game had become remedial for Junior. He was constantly searching for a new challenge. Knowing he had an impressionable and inexperienced head coach (at least on the NFL level), Junior pushed to play tight end. He'd been a standout receiver in high school and actually had taken some snaps at the position at the Pro Bowl earlier that year. But everyone knows the Pro Bowl is a free-for-all where players do things they never would do otherwise because there's no real contact in practice and players tend to look out for each other in the game.

Riley's decision was immediately second-guessed by San Diego's defensive coaches. They wondered if Riley was being manipulated by Junior's powerful and persuasive personality. To them, it made no sense to subject their best player to greater risk of injury. After all, in an experiment the previous preseason, Jason Sehorn, a starting cornerback for the Giants, tore knee ligaments returning a kickoff and missed the entire year. There also were concerns that the additional reps would take away from his effectiveness on defense.

"I guarantee you if we had the number 1 offense, this never would come up," said Junior. (The Chargers had led the league with 51 turnovers the previous season, negating their number 1–ranked defense.) "The fact here is we need an identity on offense. If I can add a spark to the offense, let's do it. I have no problems about it. We're going to put the best players on the field."

At times he looked like a credible threat, but there also were occasions when he looked out of place. Then came the preseason and a wheel route against the Chiefs. Harbaugh spotted him breaking down the sideline and floated a pass in his direction. Junior, in full sprint, reached out with both hands for the reception and carried the ball into the end zone for a 37-yard touchdown.

"I still have a picture of that in my office today," said Riley, now the head coach at Nebraska. "I personally thought it was a tremendous kind of willingness to step forward and say, 'I can help in more ways than just doing what I'm doing.' He also was just personally challenged in his career at trying to do something else. I went for it. Knowing Junior then, the physical condition that he was in, I knew his own personal pride wouldn't let him play any less at linebacker. Where else could he go to learn about the position, so it wasn't going to overburden him? And we were going to be careful about any package we used him in. It was actually kind of fun."

No one was smiling later in the year when Junior ran an underneath route for a two-yard gain and was *drilled* by Chicago linebacker Ricardo McDonald immediately after catching the pass. The hit was so violent that it opened a gash on Junior's chin.

"He got up first and celebrated like it didn't hurt, then he came to the sideline and got stitches," said teammate Orlando Ruff. "But here's the thing — he did it without the doctor numbing the area first. He was just yelling, 'Stitch me up! Stitch me up! Let's go!' They were like, 'Junior, we've got to go in the locker room.' He said, 'Fuck that! Let's go!'

"They stitched him up on the sideline, and he went out and finished the game," Ruff continued. "Craziest thing I've ever seen."

"That's maybe when I said, 'Maybe this isn't such a good idea,'" Riley said. "Maybe all the naysayers were ringing in my ears when he took that hit."

Regardless, Junior was having fun again because the Chargers won four of their first five. The joy wouldn't last, though, as the Chargers lost their next six games. Junior and the defense played a major role in the drop-off. After surrendering more than 14 points just once in the first five games, the unit allowed 28 or more five times during the skid. In four of the games, it gave up 31 or more points. Football had become a job again.

Then almost as suddenly as things went bad, the Chargers won four of their last five to finish 8-8. The defense figured prominently in the turnaround, yielding just four touchdowns in the five games. Riley also was a factor, his constant positive attitude serving as a buoy in choppy waters. At his golf tournament in the off-season, Junior called Riley to the front of the banquet room during a segment in which he presented his personal team awards for the previous season.

"It totally took me by surprise that he named me the MVP and gave me a trophy that I still have at home," Riley said. "It was him supporting what we were trying to do. His loyalty to the people around him and to me—I will just personally appreciate how he put his arms around me. Here I am, I get this unlikely job coming out of Oregon State to be the head coach of an NFL team, and I just really felt his support and his leadership and his willingness to want to help in any way. I will forever be grateful for that. He really reached out and tried to make it work. I believe Junior had a purpose for everything that he did."

At that point in his career, Junior believed that his purpose was to set the example for not only his team but also players on other clubs. He preached the importance of humility and respecting the game. He was so serious about it that he fined anyone he caught looking at themselves in the mirror during Breakfast Club workouts. He went out of his way to deliver the sermon to young talents who might one day hold the baton of greatness, like Tampa Bay linebacker Derrick Brooks.

The two first met on November 17, 1996, in San Diego Jack Murphy Stadium. Junior had already made his name as an NFL great and was heading toward his sixth consecutive Pro Bowl appearance. Brooks was a second-year inside 'backer who had yet to appear in a Pro Bowl, but he was beginning to take a claim as an up-and-comer at the position. He had size, athleticism, and innate playmaking skills that pointed to a future as someone special in the league.

"We exchanged information after that game and would talk once, twice a year," Brooks said. "The Pro Bowl is probably where I spent the most time with him. We would play dominoes together. He thought he was pretty good, I *know* I was pretty good. Our initial conversations were more about the transition at the linebacker position and what I was doing in the NFC and what he was doing in the AFC. We were trying to keep the 4-3 defense ahead of the class, in terms of the style of play.

"But he was big-talking to me about being a strong leader and passing down knowledge to guys who came behind me and played the position. He would say, 'Make sure you teach them how to take care of this league.' You listened when he spoke because he was hierarchy. That was a time when I was starting to make a run at the title, so to speak, and he was the measuring stick. We were all trying to catch up to Junior Seau."

Junior had reached a point in his career where he could relate to players on multiple levels. He was young enough to be a brother, smart enough to be a coach, and wise enough to be a grandfather. He lived in rarefied air, but never looked down on those around him. He knew the names of every teammate and their family members, from the star quarterback to the 53rd man on the 53-man roster. A figurative "welcome" mat was always at the front of his locker.

"That 53rd guy could come up and say, 'Can you help me with this?' and he never looked down on him or looked around and acted like, 'Why are you bothering me? You're not going to be around much longer anyway,'" said Orlando Ruff, who sought Junior's counsel in 1999 as an undrafted rookie out of Furman. "He would share his time. How many true stars can you say that about?"

For all the things he would and could do, Junior could not turn the Chargers into winners. He was confident they would build on the 8-8 finish in Riley's first year, but in a sign of things to come, the Chargers lost their 2000 season opener, 9–6, at Oakland. They led by four with five and a half minutes to play, but allowed the Raiders to drive 53 yards for the decisive touchdown with 2:37 to play. The following week against visiting New Orleans, they led by five with just over five minutes to play. But once again, they allowed an opponent to drive for the decisive touchdown in the final minutes, surrendering an eight-yard pass from Jeff Blake to Joe Horn with 47 seconds remaining.

One defeat was followed by another, the skid advancing from two games to four to eight to 11. In six of the losses, the defense blew a lead in the final three minutes. In four of them, they gave up the decisive score in the final 47 seconds of regulation or in overtime. This was particularly painful for a player as prideful as Junior, though he refused to show it publicly.

"No one is going to feel sorry for us," he said. "It's up to us to change things."

By this point the Chargers had become a national punch line. Leno and Letterman were using them as the butts of their late-night jokes, and nationally syndicated radio host Jim Rome was openly lobbying for them to become the first team in league history to finish 0-16. Junior shrugged publicly, but inside he seethed. He was not going to have such shame on his résumé, and in the 12th game, with the Chiefs in town, San Diego's defense played with an intensity not seen in weeks. It limited Kansas City to 10 first downs, 161 yards, and only three conversions on 18 third downs. The unit came up biggest when it mattered most—in the fourth quarter, limiting the Chiefs to a maximum of 18 yards on four possessions. The last two stops were the most important because they (a) positioned the Chargers for a go-ahead field goal from 52 yards, and (b) preserved a 17–16 decision that would be their only win of the season.

Overall, however, the Chargers presented no real fight the rest of the season, and sadly, Junior was becoming known as much for playing on bad teams as he was for being a great player. The Chargers were concluding their fifth straight nonwinning season, and in three of them they lost at least 11 of 16 games. Playing for pride was a given, but loss after loss has a way of demoralizing even the strongest-willed player. One way Junior stayed motivated was by making it personal when matched against a great opponent or an up-and-comer who wanted to challenge his de facto title of best inside linebacker in the league. Case in point: December 10, 2000.

It was a balmy afternoon in Baltimore considering the time of year. Temperature: 31 degrees. Wind: four miles per hour. Wind chill: 28 degrees. Flip-flop weather, relatively speaking.

Inside PSINet Stadium, the best and the worst of the NFL were facing off in a game that featured no anticipated drama. The host Ravens had won four in a row and would go on to win the Super Bowl. The

Chargers had lost 12 of 13 and would set a franchise record for losses by finishing a league-worst 1-15.

The only suspense was whether the Ravens would put the game away in the first quarter. The Chargers had appeared to pack it in for the season the previous week in a 45–17 loss to San Francisco. The 49ers entered that game with a 4-8 record and had been held to 16 or fewer points in three of the previous five weeks. If the Chargers couldn't do anything against them, what chance did they have against a Ravens defense that had held opponents to nine or fewer points in four of its last five outings?

The scoreboard confirmed the inevitable: 24–3, Baltimore. The Ravens forced five turnovers and limited San Diego to 128 yards on offense. And yet the outcome would have been significantly more lopsided if not for Junior, who was determined to show the sellout crowd and the rest of the NFL that the best linebacker on the field — if not the league — was wearing number 55 for San Diego and not number 52 for Baltimore.

The talk leading up to the game was that Ravens middle linebacker Ray Lewis had become the new alpha male at inside linebacker. He was younger (by six years) and smaller (six-foot-one, 240) than Junior, but possessed the same intensity, passion, and relentlessness. There was no way of missing him on the field because he was everywhere. He had that presence about him, that aura.

Junior had tremendous respect for Lewis, and their friendship was so strong that they had compared not only statistics but also contracts. When one signed a new deal, the other was the first to receive a call. Still, Junior had no intention of being upstaged by his young friend. The game was personal, not just business.

"He had the game of his life," Devaney, the Chargers' director of player personnel at the time, said of Junior, who had a game-high 16 tackles, 10 more than Lewis. "He played his ass off. Later he said, 'Ray Lewis is a great player, but I wanted to show the people in Baltimore who was the best linebacker on the field.' That's who Junior was. Ray Lewis was okay that day, but Junior was head and shoulders above him. He wasn't disparaging Ray. It was just an attitude of 'I'm the best.'"

A strong case could be made that Junior was the greatest inside line-backer in NFL history. He went to 12 straight Pro Bowls, which was a first at the position. He also was voted first-team All-Pro eight times and second-team on two occasions. He had the size of a lumberjack, the quickness of a cat, and the instincts of a clairvoyant.

Few, if any, linebackers ever blitzed the A gap (the space between the guard and the center) better than Junior. It was astounding to see a man so big get so small, often turning his body perpendicular to the line of scrimmage so he could squeeze through narrow openings. He also had a special gift for timing the snap count. He made it look so easy, but it wasn't.

"That was his thing," said quarterback Peyton Manning. "No one that I know of did it better. He could time that snap count and hit it on the dead run. He had a good feel on run/pass tendencies, and he was blowing it up on the run, taking the center and the guard, or timing it up on a pass play and making it awfully tough for that running back to block him. You have to remember, this is a *large* outside linebacker. It wasn't just an outside linebacker with speed who lacked size. He had them both. Unbelievable speed. It made him rare."

Manning faced Junior twice in his rookie season, in 1998. Junior was in his ninth year and had gone to eight straight Pro Bowls, so Manning knew he was a quality player. But his reverence climbed another rung when he listened to how offensive coordinator Tom Moore and line coach Howard Mudd spoke about Junior. They referred to him by only his first name, which was a major sign of respect. The only other player who received that first-name treatment was Buffalo Bills defensive end Bruce Smith.

Over the years Manning came to fully appreciate that respect. Manning prided himself on being prepared, on studying his opponents until he knew them better than they knew themselves. But figuring out Junior was impossible because he often ignored long-accepted practices associated with particular downs and distances. He relied on instincts and intuition as well as film study to make plays.

"You'll have an assistant coach who breaks down film of the opponent and then gives you a report," said Manning. "He'll say things like, 'This is a will [weak-side] linebacker blitz, cover-3 zone blitz,' and then you watch it and keep watching and you say, 'I don't really think that's supposed to be a blitz. Junior is blitzing on his own, but it's only because he smells something, and he's going after it.'

"It can screw up your tendency chart," Manning continued. "The will linebacker has blitzed on all these downs and distances, and you watch it and you go, 'I don't think that's a planned blitz because, if he was blitzing, the backside defensive end probably would've stopped. Those are kind of the rules of a zone blitz.' I can remember having this argument, and the assistant coach defending his work. I'm saying to him that it doesn't matter what you call, Junior will be Junior if he smells something. The amazing thing was that he was right all the time."

Junior explained it thusly: "You come into this game trying to learn. And once you learn, you understand. And once you understand, you *feel* the game."

He loved every Sunday, but particularly those that matched him against the best of the best—like in 2002 when he sat in a darkened meeting room at the Chargers complex and studied Rams running back Marshall Faulk on the projection screen in the front of the room. Faulk was as smooth as they came, a lethal threat as a runner and receiver. He had been first-team All-Pro and league "Offensive Player of the Year" in each of the previous three seasons, and in 2000 he was voted NFL MVP. Now he was trying to join Hall of Famer Earl Campbell as the only players to rush for at least 150 yards in four consecutive games.

Junior's eyes darted left and right as he watched Faulk on the screen, eager for the challenge that awaited him that weekend. "A game like this, facing a player like this, it's special," Junior told the *San Diego Union-Tribune.* "It's about competing—competing with the best, competing with a player who has been successful in his career and has longevity in the league. This is definitely a game in which you want to go after it, a game that you want to be successful and you want to help your team win.

"I'm sitting here with you talking about Marshall because, obviously, there's a lot of respect. I love talking about great players because it's hard to play this game. And to be able to play this game and compete and be successful when they have your picture in their locker room for a whole week and they're game-planning to beat you down but you can still come out there and win — that's what Marshall and I have in common. That's special. To know that an opponent is coming to take you out and is game-planning against you to allow their team to win the game? No. 1, it takes a lot of years to reach a point that you're game-planned against. But No. 2, it takes a special person to fight it year after year after year and be successful."

Even when he wanted to turn off his motor, Junior couldn't — like in the Pro Bowl, which players take less seriously than meaningless preseason games. The all-star contest prohibits blitzing, but late in the fourth quarter of one of the games Junior shot through the A gap for a tackle behind the line of scrimmage. It was a flagrant violation (though not called by the officials), and players on both teams gave him grief afterward. "He laughed and said he was fooled and lost his balance," said Derrick Brooks, cackling at the memory.

There was nothing funny, however, about the toll that losing was taking on Junior. He was the consummate professional, never publicly bad-mouthing the organization, despite its struggles. But he did make silent protests, like when he wore his shorts and sweatpants backwards. Few people knew why he did it, and when they asked he'd rattle off a one-liner that had nothing to do with anything. But once, during the 2000 season, he confided in tight ends coach Paul Chryst.

"He told me, 'Since the Super Bowl [to end the '94 season], we've been going in the wrong direction,'" Chryst said. "But he said it with a smile."

Junior's personal life was deteriorating as quickly as the team's fortunes. He was redeveloping old patterns and rekindling old friendships. The 19th Hole became a destination again. And as his drinking increased his time at home decreased. Suspicious, Gina began tracking his whereabouts with increased urgency. When he was on the road, she called his hotel room at a designated time to make sure he answered. If she couldn't reach him, she called his friends.

Privately, Junior was telling his buddies that she was obsessive and out of control. But a woman knows. She just knows.

When Gina's fears were confirmed by Junior — during her third pregnancy—the marriage was over. To friends, Junior expressed relief. He felt like an albatross had been lifted from around his neck. He was free to come and go as he pleased, as well as spend more time in Oceanside with family and friends. At least that was the public face he put on it. But privately, behind the facade of pretty, young women and wild nights on the town, he was struggling.

On the day he received the divorce papers in 2001, he phoned the house and asked to speak to Sydney. She refused to come to the phone. Ditto Jake. Suddenly he felt like someone who had jumped into the deep end of the pool without knowing how to swim. His only life preserver was a blank piece of paper, on which he spilled his soul. His writings alternated between speaking *to* himself and speaking *about* himself. He would confess his sins in one sentence, then lecture himself the next.

I totally feel lost today, more than ever. I don't know what to do with the situation. I wasn't capable of loving someone faithfully.

Then he lectured himself.

JR, you have to get on track, and it has to start with not giving yourself up in situations that will come up. You can't drink, and your friends need to change.

In the next paragraph he was back to baring his soul.

So low right now. I need to get on my feet and go after it. I need to have my family around me and hopefully get back with reality, because right now my vision is horrible. I'm not seeing clear today. I apologize for writing about nothing.

He went golfing that day but found no peace.

I couldn't think of anything except for Gina. She is on my mind every minute of my day. Whether the kids had an effect on what I'm feeling is unknown. I know one thing, and that is that I am angry about the filing of our divorce papers. I know one thing: I want my family back. I am hurting a lot of people around me and it's not cool.

Then he lectured himself again.

What do you want, because hanging around here in Del Mar is not going to work? Temptation is coming for a break point. Hopefully everything will be fine, but I know one thing, and that is that we are hurting for answers.

Junior then wrote a list of changes he needed to make:

- *I need to be honest with the person I love.*
- *I need to honor and cherish Gina.*
- *I need to be a better father to all my kids.*
- *I need to get right with my brothers and bring back my family unit.*
- *I need to change my friends and be at peace without their presence.*
- *I need to call my dad and mom more.*
- *I need to watch my alcohol intake.*
- *Rest[aurant]. Run the joint, don't drink in there or join the hostess. Bus tables and serve.*
- *I fell in love with the picture and I know that there is never going to be another Gina.*

There was no turning back for Gina. She had hired a high-powered divorce attorney and was going to make Junior pay. "The first thing he said when he told me he was going to get a divorce was, 'She's going to take me to the cleaners,'" said Annette, his sister. "Then he said, 'I can't stay. I can't stay. I know Mom and Dad are against divorce.' I told him that Mom and Dad want him to be happy. He said, 'If I don't go through with the divorce, Mom and Dad aren't getting anything. She's going to take everything.' He wanted to make sure something was left for them. I told him to do what he's got to do."

He masked his pain with more women, more liquor, and trips to the casinos. There were no outward signs of trouble because Junior was a master at compartmentalizing. He excelled at showing people what he wanted them to see, and he skillfully separated personal from professional. On the field he remained a dominant player, possessing a unique ability to troll the clubs until dawn, go straight to the training facility and sweat out the alcohol before anyone arrived, then outwork everyone on the practice field. It wasn't uncommon for him to put in a full workout before his teammates had wiped the sleep from their eyes and rolled out of bed.

But the late nights and physical play on the field were taking a toll, even if Junior refused to acknowledge it to outsiders. For instance, during training camp in 2001, he had been dominating an 11-on-11 run drill to the point that the offensive coaches became angry. Running backs coach Ollie Wilson was particularly peeved and let fullback Fred McCrary know it. McCrary was entering his third season with the team and feeling the pressure. Only a handful of jobs were safe coming off a 1-15 season, particularly with a new general manager in John Butler, who had no allegiances to anyone on the roster but those he drafted that spring.

It reflected poorly on McCrary when interior run plays failed, because he was the lead blocker on many of them. With tensions rising, quarterback Doug Flutie called for another inside run play in the huddle. McCrary felt a sense of anxiousness. Once again he would have to meet Junior in the A gap, the space between the center and the guard. He tightened his chin strap and told himself to get lower than Junior so he could win the leverage battle.

When the ball was snapped, each charged toward the other. McCrary prevailed this time, putting Junior on his back, but that wasn't what had people talking. The sound of the collision was what got everyone's attention. It was so violent that it could be heard a couple of football fields away, so violent that it left a three-inch crack across the bridge of McCrary's helmet.

"I've never seen anything like that before in my life," McCrary said afterward. "I'm going to take it home and get Junior to sign it, and I'm going to sign it, and I'm going to put it in my trophy case. That's a piece of art right there."

McCrary may have considered the helmet a badge of honor—he still does today, featuring it prominently in the downstairs trophy case at his suburban Atlanta home—but he soon realized that the damage from the hit extended far beyond the equipment. He dismissed it as a "ding" when he saw white spots after the initial collision, but as the day progressed his head began to throb painfully and he struggled with his equilibrium.

That night he ran into Junior at the dorms.

"Bug, something ain't right," he said. "My head ain't right."

"I know," Junior said. "My head is on fire! But we can't tell James [Collins, the trainer]."

They feared that Collins would keep them out of practice, so they kept their condition a secret. Months later, in the late-night silence of his bedroom, McCrary awoke shouting. He couldn't get his bed to stop spinning. It wasn't the first time it had happened, but this time seemed worse than others. He was scared, but not to the point that he heeded the words of his crying wife, who begged him to go to urgent care. He refused because he didn't want to take a chance that he might lose his starting job or be cut from the team.

No one knew how severely Junior was hurt because he never let them know. He was the leader, and as the leader he had to be there for his teammates. So he played on, keeping any pains hidden behind his smile.

12

The Trade

THE CHARGERS WERE free-falling. It was early 2003, and they had just concluded their sixth straight nonwinning season—a 5-11 finish that included nine straight losses to end the year. John Butler, the general manager who had been brought in a couple of years earlier to turn things around, wasn't accustomed to such futility.

During Butler's 14 years in Buffalo's personnel department, the Bills went to 10 playoffs and four consecutive Super Bowls. Their record was 140-83. But Butler's first season in San Diego featured no wins in November or December and no playoffs in January. Worse, coach Mike Riley had interviewed with the University of Southern California late in the year after it was apparent he would not be brought back. To Butler, a former Marine who saw active duty during the Vietnam War, that was athletic treason. He said nothing publicly, but privately he knew Riley was out at the end of the year. He had retained him in 2001 at the insistence of chairman Dean Spanos, who had promised Riley a third year on the job because there had been so much organizational dysfunction during Riley's first two seasons. But there was no saving him after a 5-2 start became a 5-11 finish.

"It came down to one bottom line, and that was [not] winning football games," Butler said after pink-slipping Riley. "Unfortunately, that didn't take place as much as we'd hoped this year, and in previous years. It just became what you knew in your heart—a change had to be made. I thought this [team] was better than a 5-11 football team. Right now the only thing I have in my mind is that I want a winner, a proven winner."

He eventually turned to Marty Schottenheimer, a respected leader who had had only one losing season in 16 years with Cleveland, Kansas City, and Washington. No coach had a better track record of quickly converting perennial losers into respected winners. In Kansas City, for instance, the Chiefs went to the playoffs in seven of his 10 seasons after doing so only once in the 17 years before he arrived. The Browns went to the postseason in each of his four full seasons, after doing so just twice in the 12 years before his first full year.

Washington was the only team Schottenheimer failed to get to the playoffs, in 2001, but that also may have been the best coaching job of his career, as he rallied the team from a 0-5 start to an 8-8 finish. Owner Dan Snyder found Schottenheimer's run-centric offense too boring, however, and he wanted the coach to relinquish some of his authority in personnel matters. Schottenheimer refused and was fired on January 13, 2002. Seventeen days later, the Chargers introduced him as their new coach.

In San Diego, Schottenheimer made his intentions clear in one of his first team meetings. He gathered the players in the large room off the lobby at their training facility and opened the session with two questions: "What's our purpose? Why are we here?"

Quarterback Drew Brees was among those seated in the front row; Junior was to his immediate left, with backup QB Doug Flutie to his immediate right. Brees considered himself a leader, but after spending his rookie season on the sideline, he had yet to start a game, and now he was reluctant to speak for the team and answer Schottenheimer.

"What's our purpose?" Schottenheimer asked again. "Why are we here?"

The coach's delivery was slow and measured, like his gait. When no one responded, his speech accelerated and his volume increased.

"What's our purpose? Why are we here?" he barked, tears welling in the corners of his eyes. If there were two things Schottenheimer was known for, besides turning losers into winners, it was for his love of hearing himself speak and for his tendency to cry as if every speech were a Hallmark moment.

Brees could see that his coach wanted an answer, but should he be the one to give it? He squirmed in his seat and glanced around the room. When Schottenheimer loudly barked the questions a fourth time, "What's our purpose?! Why are we here?!" Brees's emotions took over. He sprang to his feet and shouted a response directly in Schottenheimer's face: "*To win a championship!*"

"*You're damn right we are!*" Schottenheimer shot back.

Brees could hear his heart pounding. Reality quickly set in. What had he done? Slowly he sat down, slightly embarrassed. A few seconds later he felt a tap on the side of his thigh. It was from Junior, who gave him a nod.

"It was one of those nods like, 'That's right, young buck. And I'm with you,'" Brees said. "That was one of those moments when I knew we had a connection. I knew he was behind me."

The routine had become old hat for Junior: team loses games, new coach comes in, optimism rises. As the leader of the locker room, he knew he had to help each coach sell his program, so he would put on a positive front even when he had his doubts.

Schottenheimer simultaneously intrigued and concerned Junior. He liked the fact that Schottenheimer had a proven track record, but Junior also knew that the coach had a reputation for being inflexible and tough on veterans, even star veterans. It was said that the only thing Schottenheimer liked more than long speeches was long practices, which was something Junior never had to worry about under Riley or Jones. But Junior was willing to give the newcomer the benefit of the doubt because he was genuinely excited about the roster upgrades.

The Chargers had used the off-season to add a Pro Bowl–caliber linebacker in Donnie Edwards, a starting cornerback in rookie first-round draft choice Quentin Jammer, and a bulldozer guard in Toniu Fonoti, a second-round draft pick from Nebraska. The previous year they had drafted running back LaDainian Tomlinson and Brees with their first two picks, and Junior loved that each of these teammates was hardworking and humble.

The moves paid off early in the year. After six consecutive nonwinning seasons, the dark clouds parted as the Chargers won six of their first seven and were primed for a serious run at the division title coming off their bye week. Then they were routed at home by the Jets, 44–13, and squandered a 10-point fourth-quarter lead in a 28–24 loss at St. Louis.

They rebounded to beat the 49ers by a field goal in overtime, but the following week they were spanked 30–3 at Miami. Suddenly a sense of unease circulated through their training complex, if not the city. With three losses in four games, the Chargers were free-falling just as the Broncos were coming to town in a battle to determine first place in the AFC West.

Junior didn't practice during the week. He was battling bone spurs in an ankle that caused him to miss two of their first seven games, including a 29–6 defeat at Denver on October 6. He was able to play only by taking pregame pain-numbing injections in the ankle, but this particular week was different. The injury was worsening, to the point that no one was sure he'd be available for the game.

On Thursday night, three days before kickoff, Schottenheimer phoned him at home. "I'll never forget it because I happened to be sitting right there," said Orlando Ruff, who was at Junior's house. "Marty says, 'Look, I really need you to play on Sunday. How does your ankle feel?' Junior said, 'If you're asking me how I feel, it doesn't really matter. If you're telling me that you need me to play, I'll be out there.' From being able to barely walk during the week, he goes out and plays the entire game; not at 100 percent, but his 85 percent was better than most."

Not in this game. Junior struggled to get around and finished with just four tackles. The only positive was that the Chargers prevailed, 30–27, in overtime. But the joy was temporary, as they lost their final four games to finish 8-8 and extend their streak of nonwinning seasons to seven years.

Butler was confounded. He had fired a coach he considered to be in over his head — Riley had had losing streaks of six, 11, and nine games in each of his three seasons, respectively, and finished his tenure with a franchise-worst .292 winning percentage — and he had brought in someone who was as close to a sure thing as you could find in the NFL. He pondered what other changes he could make. Eventually his mind circled back to a conversation he had had with Spanos at the conclusion of the previous season.

"How would you feel about trading Seau?" Butler asked.

Spanos had been caught off guard. He knew they were talking about the face of the franchise and a public icon. Butler respected Junior as a player, but he knew Junior was getting older. Butler also knew that the younger players would never assume leadership roles as long as respected veterans like Junior and Harrison were around. He told Spanos they needed to do something dramatic to change the culture because it would take a year or two for the roots to take hold and produce a consistent winner.

"You have to do what you feel is best for the organization," Spanos told him. Both men knew there might be a backlash from the public. Junior wasn't a *part* of San Diego — he *was* San Diego, born and bred.

Butler had chosen not to pull the trigger the previous off-season because he believed that changing coaches *and* blowing up the roster would create too much turmoil in one off-season. After all, the Chargers still had to line up and play, and he wasn't going to find replacements better than Junior and Harrison.

But he quickly began to second-guess himself when the pair was slowed by injury. Junior missed a career-high three games, and Harrison had one of his worst years while trying to play through a torn groin that should have sidelined him for at least eight weeks; he missed just three games.

Now Butler was firm in his mind that change was needed, and he was willing to take the public backlash. He had been diagnosed with lung cancer the previous summer, and the long-term prognosis wasn't good. Feeling his health deteriorating and wanting any fallout from the move to land on his shoulders, Butler took the Chargers on a youth kick that off-season.

Never mind that in October 2000, when Junior signed a new contract extension, there was an understanding between him and Spanos that he would play only three more years, then retire after the 2003 season. Would Junior have followed through? No one knows, but it didn't matter at that point. The Chargers were moving on.

"It was a *huge* decision for all of us," said Buddy Nix, the team's top college scout at the time. "It was a deal that we labored over a lot. The bottom line was that we were rebuilding, obviously. We were losing. Junior and Rodney were the leaders of our team, and they were getting some age on them. That's not to say they couldn't play anymore; they could still play. But we felt that as long as those guys were there, players were always going to look to them for leadership.

"We wanted to develop that young group with Drew and LT [LaDainian Tomlinson] and [linebacker] Shaun Phillips. We wanted those guys to develop their own leaders. It was a *hard, hard* decision. It was one where I thought we were going to have to have protection to get in the building. Those are icons. They're not guys that you cut."

Ed McGuire, the vice president who handled contracts and financial matters on the football side, phoned Junior's agent, Marvin Demoff, and informed him of the team's plans. He then empowered Demoff to talk trade with other clubs.

Demoff had been around the game long enough to know its capricious nature, so he wasn't completely surprised. Junior had also had some forewarning.

He was hosting his annual golf tournament when his cousin Randall Godinet pulled him aside and said that something wasn't right. The Chargers had purchased a foursome, but no one had showed up. It was the first time that had happened, and their absence was conspicuous.

Later in the day, after receiving the call from Demoff, Junior sought out his cousin, who was in the bar area.

"Let's go," Junior said.

"What do you mean? Everyone is here," Godinet said.

"No, we gotta go now," Junior said.

The two then hopped in a golf cart and headed for Junior's suite at La Costa Resort and Spa. Along the way, Junior recounted what had transpired.

"My agent just got a call," he said. "He said they're trying to shop me around and said I've got 24 hours [to talk with teams]."

When they reached the room, they quickly went inside and closed the door.

"We literally stayed up the whole night," Godinet said. "Junior was crying and everything."

He had spent nearly all of his life within a 100-mile radius of his hometown, and now everything was about to change.

The next morning he met with Bette Hoffman, the director of his foundation.

"He looked at me and said, 'I'm going to be cut,'" Hoffman recalled. "I couldn't believe it. It was like a bad dream. We started putting plans together immediately on how to handle it from a PR standpoint, but he was devastated. It was like he was dead inside."

The news stayed quiet for a day or two. When it started to leak, Schottenheimer was the first to confirm it to the *San Diego Union-Tribune*, following a conversation with Junior.

"Look, this is not easy for me," the coach told the paper. "Junior Seau embodies everything you want in a player — work ethic, dedication, passion. But we felt this was something we needed to do, and we wanted to do it right. That's why we extended the courtesy to Junior and his representative to try to see if there's a trade that can be reached."

The next morning Junior held a press conference at his restaurant; no team officials joined him. Local TV stations interrupted their programming, and ESPNews carried it live. Junior wore a dark blue suit, a pale blue shirt, a red tie, and flip-flops. His face was clean-shaven, his smile frozen in place. Flanked by family and friends, he knew everyone was on edge, so he broke the ice that March 15, 2003, afternoon with a joke.

"I want everyone to know that Seau's The Restaurant won't host *The Chargers Report*," he said of the club-sponsored TV show that had aired from the restaurant the previous two years. After everyone chuckled, he got serious.

"It is true that the San Diego Chargers have advised my agent and I to . . . seek employment elsewhere," he said. "Not by choice, my agent and I are doing so. My wish here today is that I am able to go to the market and be able to seek employment without any stipulation. The support that I have received through emails, phone calls, cards, has been overwhelming, just like it has been the past 13 years. San Diego is my home and always will be. I just have to leave for six months as a professional athlete and entertain another community."

He added: "I do not want to turn this into anything negative . . . What we need to do is make sure we don't turn this into something where we're carrying a hardened heart. Junior doesn't carry a hardened heart; that's a waste of time. I have great faith that both parties will do well."

Despite the conciliatory tone, Junior seethed inside. The apostrophe in his Seau's The Restaurant logo had always been the lightning bolt found on the sides of the Chargers' helmets. But following the team's decision, he changed the apostrophe to a football. He was hurt and angered, not only by what had happened, but by how it had happened. For the better part of a decade, he had been the model of professionalism during difficult and dysfunctional times within the organization. He had never criticized the franchise or the front office in the lean years. He had never turned on teammates who weren't deserving of a roster spot. Yet he did not receive a call from the owner to personally inform him of what was taking place. He felt he deserved that, at the very least.

"We used to talk about the losing," said June Jones, the former interim coach who became a close friend of Junior's. "He was frustrated, without question, but we used to talk about how we couldn't let anything defeat us from within the locker room, because people were going to be eating away at us from the outside. We really talked about how to act, how to talk, what to say to the media, and how to deal with hard questions. The number one thing was, you always had your teammates' back and you never turned on anybody in the locker room. He was a team guy."

News of his impending departure did not go over well in San Diego.

"I've learned firsthand the NFL is a business," Harrison said. "No matter how big you are, no matter who you are, you can't escape it. Jerry Rice couldn't. Rod Woodson couldn't. Bruce Smith couldn't. Ronnie Lott couldn't. But I'm really shocked and surprised because Junior really deserves to finish his career [with the Chargers]."

"It's real sad to see Junior go," said Padres baseball star Tony Gwynn.

"He's been a big part of this community. This is what they do, unfortunately, instead of being honest with him. He deserved better than that. It's a two-way street. I'm sure he wanted to stay here, but the Chargers have to want him to stay too. For all the stuff he's done for the community, he deserved a lot better than this."

Once Junior accepted that he was no longer in the Chargers' plans, he set out to quickly find a new home. His first conversation was with Tampa Bay Bucs coach Jon Gruden, but they couldn't make it work. Then came Philadelphia. Eagles coach Andy Reid definitely was interested; he respected Junior's game and had gotten to know him during their time together at the Pro Bowl. However, Reid had committed to another player at the position and didn't want to renege on his word.

Arizona made a play for him, but Junior had no interest in the Cardinals. His focus was on the Miami Dolphins, who had gone to the playoffs every season from 1997 to 2001, though the team failed to get past the second round each time. The Dolphins were coming off a season in which they finished 9-7 and missed the playoffs altogether, creating concern within the organization that their window for winning a championship was closing.

The talent was there to make a Super Bowl run — they had seven returning Pro Bowlers, including six on defense — but veteran leadership was missing. Coach Dave Wannstedt and others within the organization thought Junior could provide it and began finalizing a restructured contract.

Behind the scenes, people who genuinely cared about Junior the person worried about his well-being. He had hit the party scene pretty hard following his divorce, and his heavy drinking was becoming more noticeable. Liba Placek, his personal trainer, noticed this. She had great respect and admiration for Junior and was concerned he might be on a destructive path. So she asked Aaron Taylor, another client, to reach out to him.

Taylor and Junior were teammates in 1998 and '99. Taylor was a burly offensive lineman who drank and partied as hard as Junior did. Sometimes they partied and drank together. Soon after retiring in

2000, Taylor recognized he had a drinking problem and sought treatment. He was open with friends and outsiders about his journey back to sobriety, and Placek thought he would be a good resource for Junior.

The former teammates, who hadn't spoken in at least a couple of years, agreed to meet for sushi in Encinitas. Their conversation began with small talk about nothing in particular. Eventually Taylor began recounting what had happened to him since he hung up his shoulder pads and helmet. A smart, outwardly confident guy, he had heard the anecdotal stories of players struggling with depression and alcoholism once they left the game. But he'd been certain, he told Junior, that it was not going to be him.

"In fact, it was me," he said.

After Taylor's confession, Junior began to open up. He said he was getting burned out on the game and didn't know how much longer he wanted to play. He mentioned the strain of being viewed as an ATM by family members and lamented the pressure of being pulled in so many directions. Taylor listened before responding.

"I told him, 'Hey, if you're struggling, or it's different than you think it might be, or you get in a little bit of a funk, become depressed, that's not unusual,'" he said. "At that time, it's very hard for athletes to acknowledge their finality. We're Supermen. We can't think about next week's game, let alone the next phase of our life. Everything has always been right here in front of us. The reason we perform at the level that we do is because we don't let anything come in and distract us."

A week or so later Junior accepted the trade to the Dolphins and signed a restructured contract.

13

Hello, South Beach

HE WAS SEATED in first-class, head tilted against the window. As the plane continued to rise while turning its nose inland, leaving behind the white caps off the coast of San Diego, Junior's stomach was a cauldron of emotions.

He was heading to Miami, where the Dolphins had scheduled a press conference to introduce him to the local media. The thought of playing on a roster loaded with championship potential excited him. The Dolphins had six players on defense alone who had participated in the previous season's Pro Bowl, including end Jason Taylor, who'd had a career-high 18.5 sacks in 2002; middle linebacker Zach Thomas, who was coming off his first 100-tackle season; and cornerback Patrick Surtain, whose six interceptions had been a personal best. Miami also had Ricky Williams, an All-Pro running back who had churned out a league-leading 1,853 yards and 16 touchdowns the previous year.

But even as the plane headed east, Junior's mind was still out west. San Diego was as comfortable to him as his favorite pair of Flojos. People had sent so many supportive emails and cards after news of the trade broke that he felt like he was attending his own funeral. Instead of focusing on the negative of how things ended with the Chargers, he thought about the good times and the love the public had shown him. Tears welled in his eyes as he silently looked down at the surf breaking along the shore.

"That was probably the most emotional time that I had during the course of the trade, because I knew my life had changed," he told the *Union-Tribune*. "I was starting a new chapter in my career, and it was either going to be bad or good. There was no gray area. Either I am going to win or I'm going to lose. To be put in that position in the latter part of my career, when all us old folks would love to just enjoy the years that we have left, that was tough. I was on that plane with a surgically repaired ankle and the insecurities of whether that ankle would be able to withstand the pressure of doing what I needed to do."

When he stepped before the media in South Florida for the first time and accepted a white Dolphins jersey, with his name and the familiar "55" he had worn since college printed in aqua and trimmed in orange, he filled the room with his smile. He gave no indication that he was conflicted inside. The mask of confidence and self-assuredness was firmly in place. It was time to go to work, and he wasn't going to show vulnerability, just as he wouldn't allow teammates and coaches to see him receive treatment when he was injured.

He dominated the press conference like a seasoned politician, stepping over potential land mines when addressing questions about his former team and how he would fit with his new club. He was acutely aware that the Dolphins had had only one losing season in the previous 26 years and were coming off six straight seasons of at least nine victories, so he talked up their history and tradition and how he needed the Dolphins more than the Dolphins needed him. He wasn't going to overstep his boundaries, so he publicly said he was comfortable with being a complementary player. That wasn't true, of course.

Junior had been a role player only once in his life—his first on-field season at USC—and he had vowed to himself it would never happen again. He was accustomed to having game plans designed to stop him, not someone else. He lived to have his picture on opponents' walls. And despite now being 34 years old, he was motivated to show that the Chargers had made a mistake. The severe ankle injury he sustained in week 3 of the previous season had limited his ability to make plays, caused him to miss three and a half games, and required off-season surgery to remove bone spurs. But now he was on the road to being fully healthy—and equally important, for the first time in his career he was going to be surrounded by elite talent.

In Junior's 13 seasons in San Diego, only seven of the Chargers' 101 defensive starters during that time reached the Pro Bowl, with Leslie

O'Neal being the only one to make more than two trips. Lee Williams, Ryan McNeil, Marcellus Wiley, and Donnie Edwards appeared in one Pro Bowl each while they were teammates with Junior; Rodney Harrison and Gill Byrd appeared in two each. O'Neal played in five during that time.

Junior felt challenged in a way he had not felt since his sophomore season at USC. This was the first time since that year that people doubted him. When he was a rookie in the NFL, everyone had assumed he could play because he was coming off a 19-sack season with the Trojans. But now he was 34, coming off an injury-slowed season. He had to make a positive impression as quickly as possible. So that off-season he increased his workouts and altered his diet, dropping 15 pounds to regain some of his quickness.

He looked the part in practice, but not in the early preseason, when people kept waiting for big plays that didn't materialize. In the regular-season opener, against Houston, he had seven tackles and no sacks or tackles for loss. The next week, against the Jets, he had five tackles and no tackles for loss. In week 3, against Buffalo, he had only four tackles, but did manage his first sack. He clearly was struggling to find his way, just as the Chargers were having problems adjusting to his absence.

They moved talented middle linebacker Donnie Edwards into Junior's weak-side spot, left second-year pro Ben Leber on the strong side, and made Zeke Moreno a full-time starter for the first time in his career, shifting him into the middle after two years as Junior's backup. The microscope they played under magnified their struggles in three areas:

1. *Chemistry:* "When you have a group of guys who've played with each other for a certain amount of years, that helps, that builds the trust," Moreno told the *San Diego Union-Tribune.* "Without even thinking about it, you know that 'I've got to do this because he's going to do that.' You're able to just worry about your job because you know the players alongside you are going to make their plays. I think when Junior and some of the [former] players were here, they had started to build that. They knew what each other was thinking and what guys were going to do. We're

just starting to build that. We thought by going away to Carson [California, for training camp] we'd build that chemistry by just working together, but it takes time. We're getting there—the trust is there. Now we've just got to keep executing our assignments."

2. *Play-calling:* "Last year we had a certain defense in which Junior was pretty much a rush guy," Edwards said. "He was very effective jumping into gaps, so we had a defense where he was able to do that. This year we really haven't run it that much. That's one of the big changes."

3. *Leadership:* "You look at it right at the outset, and you lose his energy," said Leber, a second-year starter. "I'm not looking at it as just a linebacker corps, I'm looking at it as a total defense. You lose the energy, you lose the charisma that I think no one really thought about. You were just like, 'It's just Junior,' and you went along with it. But he really pulled guys along. That's why he was such a good leader, because guys just were attracted to him and wanted to please him and emulate him. When he left, we lost a little bit of that."

Both sides began to find their legs after a month, however, setting up Junior's week 7 return to San Diego, this time as a Dolphin. He had literally been counting down to the game since the moment he signed with Miami. A digital clock in his restaurant showed the hours, minutes, and seconds until kickoff. He knew it was likely to be his only on-field chance to say good-bye to the fans who had supported him since high school.

The week started bizarrely when LaDainian Tomlinson arrived for work in a "55" Dolphins jersey to honor Junior. The move did not sit well with some people in the organization because Junior was now the "enemy," regardless of how Tomlinson felt about him personally. Tomlinson became so angry about management second-guessing his gesture that he left the building without addressing the media, a move that was out of character for him.

"What I wanted to get out of it was just paying some honor to Junior, not only [for] what he meant to me, but what he meant to the game and his dedication to the game," Tomlinson said later. "That's a lot of stuff that's missing today in football. You've got to respect people who have done a lot for the game. It's not about Junior going to Miami. It's no different from me wearing a Bo Jackson or a Joe Montana jersey. It's what they've done for the game. A lot of times people get caught up in, 'What are people going to say?' It's not about that. It's about giving the respect to a guy who has meant so much to the game and so much to me as a player."

The week got even stranger two days later when Junior wore a Chargers "21" Tomlinson jersey to work and suggested the best way to slow down the Pro Bowl runner was to feed him watermelon and fried chicken and have him "keep eating." Because Tomlinson is African American, the comment created a controversy across the sporting map. Seemingly the only ones who did not have a problem with it were the individuals involved and the friends who knew them.

Junior had that type of sense of humor. When the Chargers used him at tight end with fellow Polynesian Alfred Pupunu, he suggested the coaches call it their "Coconuts" package. When coaches would spend too much time practicing plays at the Pro Bowl, he would bark out, "Let's go, [So-and-So]. Ten-thirty tee time."

The reality was that Junior and Tomlinson ate fried chicken together on most Fridays during the season. "I'm sorry you guys took it that way," Junior said to the Miami media. "I should have realized. It was an oversight on my part. I'm sorry it came out that way."

"I wasn't offended at all," Tomlinson told San Diego reporters. "You just have to know Junior."

The sideshows soon became afterthoughts. Something far more serious was taking place. Devastating wildfires were racing through San Diego County, destroying more than 2,000 homes that sat in their path. The air was so thick with smoke — and city services were so overworked — that local officials and league executives agreed that it would be best to move the game. They settled on Sun Devil Stadium in Tempe, Arizona.

"Moving the game is very disappointing to everyone, but obviously for the safety of everyone it's probably the best decision to make," Junior said. "Right now, our prayers are with the San Diego residents and everyone that's affected by it. The game of football is secondary right now."

Junior and the Dolphins did not play that way, however. They led 10–3 after one quarter, 24–3 at the half, and 26–10 when it was over. Miami's defense was stifling, allowing San Diego to convert on just three of 14 third downs and none of its two fourth downs and permitting the Chargers to score only one touchdown on four possessions in the red zone. Surtain intercepted two passes, Thomas another, and the defense sacked Drew Brees six times. Junior had just four tackles but was presented with the game ball afterward.

"That was the hardest game I've had to play, and I think that goes for everyone, the Chargers included," Junior said. "The players there did a great job by blocking things out and coming out here and performing on the field."

When asked about playing against his former team, he said: "Those are my guys. Some of the guys I raised, some of the guys I shared a lot of years with. There's a coaching staff which I respect and an organization which I love. That's my city, that's my home, and it always will be."

Many of the same qualities that made him so special to his teammates in San Diego had carried over to Miami. He wasted no time in earning their respect. He even was allowed to address the players before breaking the huddle at the end of practice. As in San Diego, he had the guys form a tight circle and raise an arm to form a human umbrella. Then he would chant, "One-two-three," and everyone would say, "Work!"

"Leadership can't be fabricated," Junior said. "If it is fabricated and rehearsed, you can't fool the guys in the locker room. So when you talk about leadership, it comes with performance. Leadership comes with consistency. During the course of this year, hopefully everything will be where it needs to be, and when I want to be vocal or when I need to tell someone an adjustment, I can be able to do that. That is only done through trust. Trust is earned. Respect is earned."

He had earned their trust and respect, but could he get the Dolphins over the postseason hump? He struggled early in the year with a hamstring injury that caused him to miss a game, and late in the season he was slowed by a shoulder injury. In between, the Dolphins learned that playing with Junior meant they would have to adjust to him, rather than the other way around.

For instance, there was a time when he lined up in the slot and was supposed to shadow wide receiver Wayne Chrebet down the field. But Junior detected something before the snap of the ball and chose to rush the passer instead. That meant Chrebet was left to run free, which theoretically could lead to an easy score for the Jets. But Junior didn't care because he had already done the math in his head.

He knew quarterback Chad Pennington's first read would be to the opposite side of the field, away from Chrebet. He also deduced that he could get to Pennington, one of the game's savvier signal-callers, before Pennington realized what was going on, which Junior did, for an eight-yard sack.

"He'd take some really big gambles," linebacker Channing Crowder told NFL.com. "But he made 12 Pro Bowls, so clearly it worked out more often than it didn't!"

Junior's first year in Miami was solid, and overall the defense was outstanding, allowing an average of just 16.3 points a game, nearly a field goal less than the previous season and the third-fewest in the league. Unfortunately for the Dolphins, the offense wasn't nearly as good, which resulted in the Dolphins missing the playoffs despite a 10-6 record. It was the first time in 12 years that a team failed to qualify for the postseason after winning at least 10 games.

The next year Junior felt as if he were back in San Diego as dysfunction wrapped its tentacles around the organization. Williams, their star running back, retired just before the start of the season because of concerns that he would be suspended for a failed drug test. Wannstedt, the coach who helped sell Junior on the Dolphins, resigned after losing eight of nine to start the year. (The Dolphins would finish 4-12 for their first losing season since 1988.) And Junior, who was playing much better in his second season in South Florida, missed the final eight games after sustaining a torn pectoral while attempting to make a tackle on November 1 against the Jets. As if all that weren't bad enough, the Chargers were finally looking like a legitimate threat, finishing 12-4 and winning the AFC West in only their second season without him.

Junior's career was at a crossroads. He had missed more games (nine) in his last two seasons than he had in his first 13 seasons (eight). He also was going to be 36 at the start of the 2005 season. The NFL is a cruel place. Rarely is there room for a 36-year-old linebacker whose body appears to be breaking down. But new coach Nick Saban, who two years earlier had won a national title at LSU, wanted the veteran around. And Junior wanted to continue playing. The deal was consummated when he agreed to a hefty pay cut.

"I'm not here for the money," Junior said. "I'm not here for the glory. I'm not here for all the cheers and being labeled a football player. I'm here for one reason, and that's to win, to feel what goes on when you bring a whole group together and you're able to rejoice at the end of the day."

Junior was Junior when he hit the field for workouts. He was going to get in his work, but he also was going to have fun, even if it came at the expense of his new head coach. From the outside their personalities couldn't have been more different: Junior was playful, welcoming, and always looking to laugh and smile; Saban was stern, distant, and often looked as if he were suffering from heartburn.

Junior had never played for a head coach who was so tightly wound on a consistent basis, and Saban had never coached a superstar who was so consistently easygoing and jovial. Outsiders wondered how the two would mesh.

"If Junior wasn't my favorite player ever, he was close, in the top three or four," Saban said later. "The guy was a coach's dream. He had a great sense of humor, was really liked by his teammates, always worked hard. He was always pleasant, always joking around, and really enjoyed what he was doing and the role that he had on the team. That was a little bit different, to me. You always assume that when a guy is a really good player, maybe as good a player who has ever played his position, that he wouldn't be pleasant, outgoing, easygoing, and getting along with everybody. But he was. He was always busting someone's balls, especially mine.

"Once, he took the hat that I used to wear and hid it. I also used to wear these pants that had buckles on the side, and he'd say, 'Hey, Nick. Where'd you get those tuxedo pants? One size fits all?' He was always on my ass about something. Always in a really funny way, always in a really fun way. I couldn't help but crack up myself."

Junior's personality was so infectious that Saban sometimes laughed even when Junior played outside the defense. Before one practice, outside linebacker Jason Taylor warned Saban that Junior liked to make a "Baby! Baby!" call in certain situations, even though no such call existed in their playbook. The call was Junior's way of signaling that he was going to plug the B gap, so Taylor should cover for him by dropping into pass coverage and assuming Junior's responsibilities.

"I'm just telling you so that if he does that, you know what it is," Taylor said.

Sure enough, during a 9-on-7 drill, Junior saw something that told him the guard was going to pull, leaving an opening in the B gap for a split second, so he barked, "Baby! Baby!"

Saban looked at Taylor as soon as he heard the call, and Taylor broke out laughing. Junior then burst through the B gap and made the play against the running back. On the inside Saban laughed until it hurt. But outwardly he played the role of unhappy coach.

"What's this 'Baby! Baby!' shit?" he said to Junior.

"I knew it was a power [run], and I knew if I ran to it I could make the play," Junior responded.

"He was serious as hell," Saban said. "I'm crying now, laughing so hard, telling you these stories."

Junior could get away with these antics because everyone knew he was committed to chasing a championship. Teammates had voted him the winner of their leadership award each of his first two seasons, and now, with his 16th season approaching, he still was moving with the urgency of an undrafted rookie trying to earn a roster spot.

"He was the one who set the example—full gas, full out, wide open all the time," says Seattle Seahawks defensive coordinator Dan Quinn, who was an assistant that year in Miami. "You see him and you're like, 'That's it. That's balling. That's how you practice. That's what it feels like and looks like.' It was kind of cool. The guys that have those kinds of practice habits, as a coach, you appreciate that. Not everybody has that.

"Once, he came in for OTAs [off-season training activities] after taking the red eye from San Diego and went straight to the workout. He took every rep on defense, and when we went to the cards he wanted all the scout team reps to show the guys on offense that I'll do anything for you. We're talking about a 12-time Pro Bowler who was nearing the end of his career. He had earned the right not to do those things, yet it didn't matter to him. That guy got people. He recognized how to connect with his teammates. He was a real guy and was held in such high regard. Certain guys got it. He was one of them."

Unfortunately for Junior, his body would betray him again. Slowed by Achilles' tendon and calf injuries throughout the year, he was placed on injured reserve in November, ending his season and raising questions about whether he would ever play again. He was under contract for another season, with a salary and bonuses that totaled $2.1 million, but Miami wasn't going to honor the deal. Saban already was trying to infuse the roster with more youth, and Junior was now three years removed from the last time he played all 16 games in a season. The curtain was falling.

14

The Graduation

FOR THE FIRST time in his professional life, Junior was on the street. Even at 37 years old, he still felt he could play at a high level, but he needed for someone to *need* him rather than want him. It was a subtle yet significant difference.

He waited for the right phone call, but when March morphed into April . . . and April turned into May . . . and May lapsed into June and still he'd received no credible call, Junior sensed he was staring at the finish line. For the first time he began to seriously consider life without the Sunday spotlight. Taylor, his former teammate in San Diego, recognized this and reached out to him, just as he had done three years earlier before Junior was traded to Miami.

The two met at an Italian restaurant in Oceanside. Taylor quickly realized that while things were changing in Junior's professional life, the former All-Pro was still going as hard as ever in his personal life with the drinking and partying.

"I brought him a book from a 12-step alcohol rehab program and just wrote something to him inside of it," Taylor said. "I didn't think he was ready, even though he indicated he was, but I could sense he was curious. He just wasn't ready to really jump in, so I gave him that book and that was about it."

Typical of their relationship, Junior went off the grid after that get-together. They'd text each other and agree to have dinner or lunch, but Junior wouldn't respond when Taylor tried to finalize a place or time.

Another person reaching out to Junior was Jim Steeg, though for different reasons. Steeg was the executive vice president and chief operating officer of the Chargers, and he wanted Junior to end his career as a ceremonial Charger. So shortly after Junior was cut by the Dolphins, Steeg phoned Bette Hoffman, the director of Junior's foundation, and expressed interest in having a retirement ceremony at Chargers Park, during which Junior would sign a one-day contract.

Steeg, who had joined the franchise in 2004, was disturbed that a number of marquee alumni had distanced themselves from the organization after leaving it. The list included tight end Kellen Winslow, who resented being portrayed as a malingerer at the end of his career while trying to overcome an injury; defensive end Fred Dean, who was traded to San Francisco because the team didn't want to pay him like the dominant player that he was; wide receiver John Jefferson, the most popular player on the roster, who was dealt to Green Bay after he held out for more money; safety Rodney Harrison, who was released and characterized as old and unable to run after playing most of his final season in San Diego with a 30 percent tear of the groin; and all-star quarterback Jack Kemp, who was claimed by Buffalo for a paltry $100 because the Chargers didn't know the waiver rules. Offensive tackle Ron Mix left on good terms, but former owner Gene Klein unretired Mix's jersey number when Mix returned to the league and signed with the Raiders, a hated division rival.

Steeg believed that honoring Junior might serve as balm for whatever wounds were still open. When he got the call from Hoffman that Junior was likely to retire, Steeg began formulating a plan to bring home the native son. He even contacted some of the franchise's past greats and asked them to attend the news conference. The Chargers also brought in kids from the Oceanside Boys & Girls Club, which was close to Junior's heart because he had spent countless hours there as a youngster and felt a kinship with the kids.

"All Junior wanted to do was help the poor kids in Oceanside," said Mary, his sister.

Ronnie Lott saw this in Junior as well. Lott was an All-America safety at USC who went on to become a Hall of Fame player. He and Junior grew close over the years because of their USC allegiance and the fact that they played the game the same way: all out. Over the years he was struck by Junior's intensity on the field and compassion off of it.

"We all break down when we see something that's not right in life, and growing up the way he grew up — there were a lot of things that he knew weren't right for his people and for his community and the kids in it," said Lott. "We would talk about it, and he was emotional that kids didn't get certain things. He was always in pursuit of helping somebody have something that he didn't have at that age. That brought a lot of joy to him, and at the same time probably brought some pain — the pain of just acknowledging the things that he went through growing up. It was tough."

When Junior arrived at the team's Murphy Canyon training facility for the ceremony, the weather was hot and windy. He stopped and spoke to the kids from his hometown, and then he delivered a passionate speech to the Chargers players after they concluded practice. He worked himself into such a lather that the club had to give him a pair of shorts and a T-shirt so that his suit and dress shirt could be put in a dryer before the ceremony began.

Junior hung out in team president Dean Spanos's office in the meantime. It was their first serious discussion since he had been traded, and it felt good to close the door on some of the painful past. The two then headed downstairs, where the press conference was to be held outdoors beneath a large white tent, with 400 guests looking on. There were blue-and-gold tablecloths, matching balloons, and hugs and smiles everywhere.

When Junior took the mic, it was as if church was now in session. He didn't give a speech as much as he gave a sermon. "Live for today, work for tomorrow, pray for the rest," he told everyone. He tried to wipe away the sweat on his face with a thick white towel, but it didn't do any good. Within no time he had sweated through his dress shirt and suit jacket again.

The media peppered him with questions about retirement, but he refused to use the word, saying instead that he was "graduating" to the next phase of his life. He even declined to identify what that phase would be, though he claimed to be ready for it.

But later that night, at a small private party at his restaurant, he revealed to a handful of close friends that a comeback might be in the works. He was seated at a patio table outside his restaurant. Music flowed through the speaker system, and the moon provided much of the light.

"The Patriots called," he said.

"It had been in the works for 24 hours maybe," said Marvin Demoff, his agent. "Junior waited and waited because he wanted to play, but the phone wasn't ringing, and I had too much respect for him to call 31 other teams and say, 'Do you want Junior Seau?' I said, 'Let's wait and see if there's something out there.' Then there was a call. Scott Pioli called me."

Pioli was a senior executive with the Patriots and the right hand to head coach Bill Belichick.

"There was a linebacker that the Patriots thought they were going to get from the Dallas Cowboys that never transpired," Demoff said. "So they called me about Junior. The Patriots — like any team when they want someone — were pretty good sellers. We talked about Junior coming there, and Junior talked to Belichick the day before the press conference. Junior went through with it because there was no guarantee a deal would get done."

For a while it appeared like the talks would collapse. Pioli was offering less than what Junior was seeking. And though the difference was "significantly" less than $100,000, Junior was firm on the number he wanted. That was when Belichick stepped in, agreed to the figure, and closed the deal.

"I always got a kick out of Belichick wanting to be the kingmaker," Demoff said. "Junior felt great because Bill really wanted him."

"There were teams that wanted me, maybe at leadership for the locker room, or a plug-in linebacker, sparingly, but there wasn't a team that needed me," Junior said. "A call came, and it was Bill Belichick, and the words he used on the phone were the reason why I'm here. He said to me, 'I have a position for you. This year.' Those were his words. That's a team, a coach, that's a locker room that needed my service, and I was willing to do it."

The Patriots had won three of the previous five Super Bowls and were poised to make another run with some tweaking and good health. But first they needed to upgrade at wide receiver, where former Super Bowl MVP Deion Branch was holding out in a contract dispute. They also needed to add depth at inside linebacker, where injuries and ineffective play had become a concern. Starter Tedy Bruschi was sidelined with a wrist injury, and no one had stepped in to fill the void. In fact, the Pats contacted Demoff the week after being gashed for an average of nearly six yards a carry and 196 yards rushing overall in a preseason-opening loss to the Atlanta Falcons.

Belichick knew a healthy Junior could help. At six feet, three inches, and 250 pounds, the savvy veteran possessed the size, temperament, and instincts to slow opposing ball-carriers. Plus, Belichick had always admired Junior's game from afar. In 1994, while coaching the Browns, he told reporters that Junior was "the best defensive player we've faced — I'd say by a pretty good margin."

Four days after his "graduation" ceremony, Junior signed a contract with the Patriots. As much as he tried to convince himself he was prepared for retirement, he wasn't. There was nothing that could fill the competitive void — or pay as handsomely.

The Patriots had provided him with an unexpected gift when they called, because now he would have a chance to fill the gaping hole on his résumé: no championships. The Patriots were so good that they didn't hope to win; they *expected* to win. There was an athletic arrogance about them that Junior admired and respected. It had now been 11 years since he last appeared in a playoff.

"We recognized that he fit in right away," said Bruschi, who was sidelined when Junior arrived yet played in all but one game during the season. "I feel like he had never experienced a program like ours before — or a group or a team like ours before. He would always talk about the amazing humility on the team. You couldn't have a locker room that was more accomplished; most of us had won three Super Bowls. But we felt like none of our roles was more important than the guy next to us. Junior recognized that right away."

Dean Pees was in his first year as New England's defensive coordinator when Junior arrived. The two instantly connected beyond football because Junior was older than many of the other players and therefore could discuss certain life experiences with Pees, who had 20 years on his new pupil. Junior also was just learning to play the ukulele, and Pees was a guitar man, so they shared a common interest in music.

Junior was not physically dominant at this point in his career, but he made up for the lost step with a high football IQ. That knowledge was a double-edged sword, though, because while it allowed him to be productive, it also led him sometimes to be out of position because he'd attempt to make plays outside of the defense.

On one particular third-and-long situation, Pees called for a three-deep zone with a four-man rush. Junior, the middle linebacker, was supposed to drop 10 to 12 yards and cover the hook zone. Instead, he rushed the passer and got the sack.

When the play ended, Belichick turned to Pees and asked what play he had called. Pees made up a name to protect Junior. He called it "Mike Robber," or something to that effect. Junior was laughing as he came off the field.

"You're killing me. You're killing me," Pees said, laughing himself.

Over time the two came to an understanding: Pees stopped making calls that required Junior to drop into coverage, instead allowing him to improvise along the line as the fifth rusher in passing situations. In return, Junior agreed to be more disciplined in other situations.

"It was kind of like, 'You give me this, and I'll give you that,'" said Pees. "I used to call him 'gangster,' because he was known to change the defense from the call I made during the game. But we kind of negotiated a deal where, 'I'll give you enough freelance plays if you give me enough discipline plays and be where I need you to be.' It worked out great.

"If it's a rookie who improvises, that's a different deal. You haven't earned the right. But with Junior, there was always some logic to what he was doing. He knew something from watching film, or he saw something that was a giveaway. It was never one of those things where, 'I'm just going to do what I want to do.' He was doing it because he felt like he knew what he was doing and it was going to help the play. Once in a while it didn't, but you've got to live with those. I loved the guy."

Junior did not perform like someone ready for retirement. He started 10 of 11 games and led the team in tackles in three of them. In another game, he finished second in tackles. Equally important was that the Patriots held all but two of their opponents to less than 100 yards rushing in games he started and finished. He and Bruschi complemented each other well because they were experienced—Junior was in his 17th season, Bruschi in his 11th—and both had intelligence to go with their instincts. That allowed them to adjust to each other on the fly.

"There were a lot of times I knew Junior would take chances," Bruschi said. "He would tell me he felt something or saw something, and I would give him the freedom to keep playing the way he was playing because I knew the defense well enough to where I could make a call to adjust a defensive lineman to what Junior wanted to do, or I could adjust my alignment or responsibility based on what he was going to do. For instance, if I'm a linebacker in a 3-4 defense on the right side, and I know Junior is going to attack the A gap because he feels it's a backfield set that he should attack, and I see him cross my face to the A gap, I know that if the play goes to the left he's now backside and I need to go over a couple of gaps because he has that A gap covered. He would say, 'Broo Broo, cover me,' and I knew what he meant. I'd play off of him, and I had no problem doing it because I really respected him as one of the great linebackers of my generation."

Then it happened. Again.

With just under nine minutes to play in the second quarter of a November 26 game against visiting Chicago, Junior dove low and attempted to wrap the legs of running back Cedric Benson from behind. When he landed awkwardly on his right arm, the radius and ulna bones in his forearm snapped, ending his season. It was a major disappointment for Junior and a considerable loss for the Patriots' run defense, which surrendered a season-high 153 yards that day and allowed 100 yards or more in four of its final five games. In fact, the Patriots' last three opponents averaged 4.8, 6.5, and 5.4 yards per rush; the league average was 4.1 that year.

The injury was gruesome. His arm was broken cleanly just below the midpoint of the forearm, leaving it hanging limp at a near-90-degree angle. He used his left hand to hold the lower half of his arm level with the upper half. His teammates were his first thought when he realized what had happened. He didn't want them to see the injury, so he curled into a ball on his knees, in obvious pain, and held his arm against his torso until the training staff arrived.

As he walked off the field to cheers, he waved to the crowd with his left hand. He didn't know if it would be the final game of his career, and he wanted to acknowledge the respect being directed at him. He flew home to San Diego a day or two later and had screws surgically inserted into his arm. Then, typically for Junior, he was back working out a week later, this time going through grueling private workouts with members of the San Diego Padres, among them Trevor Hoffman, Dave Roberts, and Mike Sweeney, who were preparing for spring training.

Liba Placek was the personal trainer for the group. She looked at Junior and his surgically repaired arm and shook her head, as if to say, *What are you doing?* Junior just smiled in return.

"He'd go through the entire workout and say, 'What? I can do everything on my elbows instead of my hands'—and he did," Placek said. "When we went to play volleyball afterward, I told him he couldn't play because of his arm. He said, 'What? It has screws in it. It's much more durable than before.'"

Everyone laughed, but the things Junior put himself through were no joke.

"People always said, 'Does this guy never get hurt?'" former teammate Orlando Ruff said. "It was just the opposite—he got hurt all the time. But what would make you and I stand down or sit out, that wasn't an option for him."

Junior was determined to continue his career. He had shown that he could still play at a high level for a quality opponent, and he wanted another opportunity. Very quickly it appeared as if 2007 would provide him with the fairy-tale ending he so desperately desired.

The Patriots had it all that year: a big, physical defense that played with cohesion and tenacity, and an offense that would break the league record for points scored, thanks in large part to trades for wide receivers Randy Moss and Wes Welker.

Moss, a talented yet temperamental veteran, was the game's premier deep threat for most of his first eight seasons in the league, but the Patriots acquired him for a cup of coffee from the Oakland Raiders because his production and his attitude had gone south. So far south that the Raiders all but gave Moss away two years after sending starting linebacker Napoleon Harris and first- and seventh-round draft picks to the Minnesota Vikings to acquire his rights. He had had a decent season in 2005 with Oakland, but the man who once said, "I play when I want to play," set career lows for catches (42), yards (553), and touchdowns (three) the following year. His speed appeared to have gone from 12 cylinders to six, and his petulance was more than owner Al Davis wanted to deal with. So he sent Moss to New England for a paltry seventh-round pick.

In a separate move, the Patriots also acquired Wes Welker from the Miami Dolphins for second- and seventh-round picks. New England envisioned Welker, a slithery slot receiver, beating one-on-one match-ups underneath when Moss commanded attention down the field. The strategy proved to be genius: one year after throwing for 3,590 yards and 25 touchdowns, the Patriots' passing game generated 4,859 yards and 50 touchdowns.

Capitalizing on the brilliance of quarterback Tom Brady and the creativeness of coordinator Josh McDaniels, Moss caught 98 passes for 1,493 yards and 23 touchdowns. The yards ranked second in his career, and the touchdowns broke Jerry Rice's 20-year-old record. Moss's ability to stretch defenses created underneath opportunities for Welker, who had 112 receptions for 1,175 yards and eight touchdowns. The Patriots would go on to score an NFL-record 589 points and set the all-time modern-era mark for largest scoring differential in a season, at 19.7 points. They became the first team to go undefeated in the regular season since the schedule expanded to 16 games in 1978, and they entered the playoffs as the unquestioned favorites to win the Vince Lombardi Trophy.

They got an unexpected tussle from the Jacksonville Jaguars in their playoff opener, then prepared to face the Chargers for the right to advance to the Super Bowl. Surely the football gods couldn't be so cruel as to allow Junior to get this far—healthy, no less—only to fall to his former team, whose quarterback, Philip Rivers, was playing on a knee that had been surgically repaired earlier in the week and whose star running back, Junior's buddy LaDainian Tomlinson, would go to the sideline early in the first quarter with a knee injury and not return.

For a time the improbable seemed possible. The Chargers refused to blink at their injuries, the cold (with wind chill, nine degrees), or their favored opponent. They led 3–0 after one quarter and trailed just 14–9 at the half. Momentum appeared to swing dramatically in their favor to start the third quarter when cornerback Drayton Florence picked off Brady near midfield and returned the ball to the Patriots' 49. Seven plays later, the Chargers were at the New England 4-yard line. They needed only a yard for a first down.

A current of concern raced through the sellout crowd, which wondered how this was happening. A touchdown would give the Chargers the lead and confirm that they were destiny's darlings, having reached the conference final one year after firing a coach (Schottenheimer) at the end of a 14-2 season. Someone needed to make a play.

As he had done for much of his career, Junior correctly deciphered that the right guard was leaning to his left, indicating that he planned to pull in that direction. Junior blitzed from the backside and tackled running back Michael Turner for a loss that forced San Diego to kick a field goal. So, instead of taking the lead, the Chargers trailed 14–12 and never got closer in a 21–12 loss.

"Most of those weren't called," Belichick said of the blitzes. "Believe me, Junior and I talked a lot about that. He just took it instinctively, based on the snap count and sometimes the split of the linemen. If he hit it, he hit it. If he didn't, he would just kind of bounce out of there and go back to his assignment, which was usually the hook zone in the passing game. He was so disruptive with those plays."

New York Giants Hall of Famer Lawrence Taylor is the only other linebacker Belichick puts in Junior's class when it comes to having similar instincts, explosiveness, and timing on the blitz.

"Junior and LT both had sort of the same mentality," he said. "Honestly, I don't think either of them felt there was a player on the field who could legitimately block them . . . It didn't matter whether it was a run or pass or play-action, Junior was going to go in there and blow it up before it got started. He could do that three, four, five times a game, and do it very effectively. *There's too much space in there, I've got too much momentum, I've got too good of a shot at this, they can't get me.*"

And yet, for all the fairy-tale angles following the win over the

Chargers, it turned out that the football gods are not fans of happy endings. They allowed the Patriots to get so close to immortality—the first 19-0 finish in league history—only to have them squander a lead in the final minute of Super Bowl XLII and lose 17–14 to the New York Giants.

The Pats were up 14–10 with 2:42 to play after Randy Moss caught a six-yard touchdown pass from Brady to complete a 12-play, 80-yard drive. But the Giants answered with an incredible 83-yard touchdown march, highlighted by a 32-yard pass play that began with quarterback Eli Manning making like David Copperfield to escape the pass rush, then ended with wideout David Tyree pinning the ball against his helmet with one hand while being pulled to the turf by safety Rodney Harrison. Four plays later, Manning beat an all-out blitz to find Plaxico Burros in the corner of the end zone for the decisive 13-yard touchdown.

"I'm still heartbroken about that last play," said Belichick. "I can still see Junior laying on the field, face down, after Plaxico caught that pass, kind of like his chance at a title had slipped away. It was a heartbreaking moment in many respects, but especially for him and his career. That's all we talked about—winning a championship. That's why he came back and played. It wasn't about the money. He came back and played because he wanted a championship."

"It's still one of the biggest regrets of my career," Bruschi said. "Not missing out on going 19-0, but remembering seeing Junior on the field, and remembering that we couldn't get it done for him. That would've been so special. That 19-0, that's what was on everyone's mind and lips. But the deeper meaning of it would be getting one of the best linebackers in the history of the NFL his championship too."

Unlike in his previous loss in a Super Bowl, this one stayed with Junior for months. He could accept the 23-point spanking in Super Bowl XXIX because the 49ers were clearly the superior team. Stacked at nearly every position, San Francisco had led 28–10 at the half and 42–18 after three quarters. The loss to the Giants was different, though. One more defensive stop and he would have had his first championship and the immortality associated with a perfect season.

When he ran into a friend in the interview area outside the locker room, he smiled and shrugged, trying to keep his spirits up. He didn't say anything, but his body language seemed to say: *What are you going to do?*

The Patriots chose not to immediately re-sign him the following year, but the sides were reunited in mid-October when the team traveled to San Diego to face the Chargers. While there, owner Robert Kraft arranged for a special ceremony at Junior's restaurant to present him with an AFC championship ring from the previous season. Junior did not know it, but it wouldn't be the last time he'd see the Patriots.

Two months later, on December 5, Belichick came calling again after a string of injuries at inside linebacker left the Patriots thin at the position. Junior had tremendous respect for Belichick, so he signed what amounted to a one-month contract. The Patriots were still in the hunt for the playoffs, but Junior knew the chances of them winning a title were slim to none. New England had lost Brady, its All-Pro QB, to a season-ending knee injury in week 1; Matt Cassel was a solid backup, but he was no Brady.

Ultimately, it proved to be a moot point because New England failed to qualify for the playoffs despite winning 11 games. Junior found a seat on the first plane out of town, eager to return to the sun and surf of Oceanside—if not to get on with the rest of his life. It was time. But his tranquillity was shattered a month later when he received a phone call telling him that his close friend Mike Whitmarsh had committed suicide.

Like Junior, Whitmarsh grew up in San Diego County and made a national name for himself in college, earning honorable mention All-America honors as a basketball player at the University of San Diego. As a senior in 1984, he led the Toreros to their first West Coast Conference championship—not to mention their first NCAA tournament appearance—by pacing them in scoring (18.8 points), rebounding (7.3), and assists (6.0), something no other player has ever done in school history.

When he failed to make it in professional basketball, Whitmarsh took his six-foot-seven frame, gravity-defying leaping ability, and all-around athleticism to the sand courts of beach volleyball, where he would establish himself as one of the sport's all-time greats. In fact, he won an Olympic silver medal and 29 tour titles over a 15-year career, becoming a face of the sport along the way.

To those looking in from the outside, he had it all: a wife and two children, the respect and adulation of fans and opponents, and the love and support of family and friends. But on February 17, 2009, he was found dead in a friend's Solana Beach garage. The medical examiner listed the death as suicide from inhalation of carbon monoxide, produced by a running car in an enclosed area.

The report stunned everyone. Whitmarsh was known to be a great guy who was personable and kind. He didn't appear to have a care in the world. It was learned later that he had recently signed divorce papers, but was that enough for him to take his life?

Junior was particularly shaken. He and Whitmarsh had hung out together, drunk together, partied together. They could relate to each other in ways that others could not. Junior agreed to speak at Whitmarsh's service, then afterward held a "celebration of life" for him at his restaurant.

That afternoon, during a quiet moment with some members of his inner circle, Junior said: "We need to make a pact. If any one of us has a problem, just come and reach out to any of us, and we'll be here for you. We're not going to judge you. We're just going to love you."

Everyone agreed, and with football in his rearview mirror, Junior sought new ways to fill his time. One involved shooting a TV pilot called *Sports Jobs with Junior Seau.*

"After 19 years of pro football, I thought I'd done it all," he said in the promotional trailer. "But I never had to close a three-inch cut with one hand. I never needed to change four tires in under 10 seconds. And I never stopped a 100-mile-per-hour line drive. Until now."

The show chronicled him doing behind-the-scenes jobs like working as a corner man for a mixed-martial artist, participating on an Indy car pit crew, spending a day as a sportswriter, and working as a rodeo clown. In fact, he returned to Boston, in late September, as the equipment manager for the Washington Capitals, who were taking on the Boston Bruins in both teams' NHL opener.

His stopover generated obvious questions about whether he would make another return to New England to re-sign. Now 40, Junior side-stepped the matter. Belichick did not, however.

During his weekly radio appearance on WEEI, Belichick said that Junior had undergone a physical with the team and might sign with them. Typical of their close relationship, Belichick couldn't help but have some fun at the linebacker's expense, mentioning that Junior's stint as a rodeo clown on the TV pilot had not gone well.

"I noticed he was doing some bull riding or bull stomping or bulls were stomping him or whatever it was," said Belichick. "We'll have to take a look at that workout and see how he was doing that."

Junior, who had taunted a bull by getting in a three-point stance, was promptly run over during a stop on the Professional Bull Riders Tour. Sure enough, he was back on safer ground that October after signing with the Patriots.

His 20th NFL season would turn out to be unlike any he had experienced. In 2009, for the first time, he was wanted but not needed. He appeared in only seven games—none as a starter—and totaled just 14 tackles, which was less than he'd had in some games during his heyday. In a playoff-opening 33–14 loss to the visiting Baltimore Ravens, he recorded just five tackles, just one coming in the second half.

As he stripped off his uniform in the locker room, he knew he was doing it for the final time. His career had reached the finish line. The team knew it. The fans knew it. Most importantly, he knew it.

"His last year in New England, he didn't play much the last half-dozen games, and he felt abandoned," said Demoff, his agent. "I think that his lack of contributions, the lack of being needed, just the way his career ended, made him feel very hollow. I don't think he ever addressed it or dealt with it. I think he really felt like the NFL abandoned him at the end. You understand that there are fires and that houses burn down, but they're not your house, so the impact isn't as great. I think that's what happened to him; he realized his career was over and he didn't get to go out on his terms."

15

"This Is Not Who I Want to Be"

ON MARCH 2, 2012, NFL officials accused the New Orleans Saints of running a program that paid illegal bonuses to their players for injuring or knocking opponents from games. The allegation shook the league to its core. It already was on its heels from being sued for allegedly hiding the long-term dangers of concussions from players, so the notion that Saints coaches and management were directing and sanctioning a "bounty" program that paid out $1,500 for a "knockout" hit and $1,000 for a "cart-off" hit was both a PR nightmare and a potential legal land mine — particularly at a time when Commissioner Roger Goodell was pushing for increased enforcement of player safety rules.

Within locker rooms, players scoffed at the suggestion of a pay-to-injure program. NFL players believe they are part of a brotherhood, and while they willfully and violently attack each other on Sundays, the idea of intentionally jeopardizing a family member's livelihood for a few extra dollars was revolting to most of them. Players were known to pool monies and reward teammates for big hits and big plays, such as sacks, interceptions, and forced fumbles. But that was a far cry from compensating someone for attempting to — and succeeding at — injuring an opponent.

Junior was two and a half years into retirement when the scandal broke. The suggestion that a player would intentionally seek to injure an opponent caused him to shake his head.

"That's some underground market they've got going on if it's true," he said. "I've never heard of anything like that. If it's true, it's the wrong thing to do. The way I played the game was to inflict pain on my opponent and have him quit. It was never to get paid for getting him out of the game. You should never incentivize anyone's health in the game of football. That is wrong. But to strike your will on another player in hopes that the player quits on you and allows you to do what you need to do at your pace—that's the name of the game, to have your guy surrender. And once he surrenders, you don't stomp on him; you go on to the next guy."

Imposing his will often meant jeopardizing his health.

For instance, in a game late in his career with the Patriots, he broke his hand making a tackle near the end of the second quarter. At half-time he was sitting in the locker room with a bag of ice on it when a trainer approached.

"We need to go for an X-ray," the trainer said.

"Why?" Junior said, then added with a chuckle, "I know it's broke. We don't need to X-ray it."

Then he went out and played the second half without complaining or making an excuse.

Playing through pain is expected in the NFL, where the men are so big and fast, and the collisions so fierce, that it is easy to believe, as Arizona quarterback Carson Palmer once told *Sports Illustrated*, that "somebody is going to die" on the field. More frightening than the players' willingness to play through pain is the frequency with which they're willing to play through injuries, even if they know long-term consequences are possible.

In 2006, Washington Redskins outside linebacker/defensive end Jason Taylor was leg-whipped by an opponent and sustained a deep and painful calf bruise. He didn't know the true severity of the injury because it was masked by the Toradol injection he received before the game and the pain meds and prescription sleeping pills he took after it. Reality set in at 2:00 AM when the drugs wore off.

"He noticed that the only time his calf didn't hurt was when he was walking around his house or standing," Dan Le Betard wrote in the *Miami Herald*. "So he found a spot that gave him relief on a staircase and fell asleep standing up, leaning against the wall. But as soon as his leg would relax from the sleep, the pain would wake him up again. He called the team trainer and asked if he could take another Vicodin. The trainer said absolutely not . . . The trainer rushed to Taylor's house. Taylor thought he was overreacting. The trainer told him they were immediately going to the hospital. A test kit came out. Taylor's blood pressure was so high that the doctors thought the test kit was faulty. Another test. Same crazy numbers. Doctors demanded immediate surgery. Taylor said absolutely not, that he wanted to call his wife and his agent and the famed Dr. James Andrews for a second opinion. Andrews also recommended surgery, and fast. Taylor said, fine, he'd fly out in owner Daniel Snyder's private jet in the morning. Andrews said that was fine but that he'd have to cut off Taylor's leg upon arrival. Taylor thought he was joking. Andrews wasn't. Compartment syndrome. Muscle bleeds into the cavity, causing nerve damage. Two more hours, and Taylor would have had one fewer leg.

When asked his reaction to the news, Taylor told Le Betard: "I was mad because I had to sit out three weeks. I was hot. Players play. It is who we are. We always think we can overcome."

Steve Wisniewski, the Oakland Raiders guard who had fierce battles with Junior, was one of the league's orneriest players during a 13-year career that ended after the 2001 season. He refused to accept defeat from an opponent or an injury. During a game at Buffalo he suffered a blow to the head that prevented him from standing upright on his own. He refused to leave the field, though, relying on teammates to hold him up until they broke the huddle for the next play.

"When Jeff Hostetler, our quarterback, called the play, I would turn to a buddy and ask, 'Is that a run or a pass?'" Wisniewski said. "I completely blanked on the scheme. But it was a different culture then. In that time period, you were really looked down upon if you couldn't be out there. If you were injured, you found a way to play. I missed one game in my NFL career. A second game I was dressed and my coach wouldn't let me play because I was literally hobbling. A great many times you're in the locker room wondering, *How am I going to get through this game?*

"I liken it to the military. I haven't served in the military, but my father, my brother, my son, my brother-in-law, my uncle — they all served. In the military guys do extraordinary feats of heroism because of their buddies, for the people in their unit. They don't do it for heroism or country or politicians. They do it for each other. When I played I always had that sense that you're not going to let your team down. Junior was that way. He pushed himself. He was there for the team."

The desire to be accountable sometimes causes players to self-medicate. Although the behavior is dangerous and potentially life-threatening, the risks are considered acceptable to some players because the rewards — real and perceived — are so great.

How do the players get the drugs? First, they hoard team-issued painkillers over time for fear they won't be able to get them from the club at a later date. Then some will barter the unused meds for stronger prescription drugs from teammates.

Say you sprain an ankle and receive 10 Vicodin from the club. If you have a high pain threshold, you might use only three and store the rest in the medicine cabinet. A few weeks later you break a finger and are given 10 more Vicodin, of which you use only five. Intentional or not, you now have an excess of powerful pills to use when and how you please, with no medical supervision.

"There were some practices that you could consider abusive when it came to the use of prescription meds," said Terrell Fletcher, a running back who played in 111 games for the Chargers from 1995 to 2002. "Fortunately for me, I did not see any team officials assist with that, but it's not hard to do. People are in pain, guys want to get on the field and play, and sometimes you have to do what you have to do to try to keep yourself on the football field."

Each team's medical staff is required by law to log the painkillers that are administered to players, but there's no way for them to ensure that the pills are taken in full or as prescribed. There's also the issue of players who lie about the severity of their injuries, claiming after the prescription has lapsed that they're still in pain so they can stockpile meds for future unsupervised use.

"Self-medicating is very scary, but it does happen," said Fletcher. "When you're given anything unlegislated, it can be dangerous."

The NFL constantly contends with the perception that there's a conflict of interest when players are treated by doctors paid by the club. The thinking is that clubs need players on the field to win games and sell tickets, so they're going to do what's in the best interest of the franchise short-term rather than what's in the best interest of the player long-term.

In 2014 more than 750 retired players filed suit against the NFL, alleging that it put them at risk by routinely and illegally providing them with prescription pills and painkillers to keep them on the field. The 87-page complaint, filed in US District Court in San Francisco, states that the NFL "has intentionally, recklessly and negligently created and maintained a culture of drug misuse, substituting players' health for profit."

Marcellus Wiley, a defensive end who played for Buffalo, San Diego, Dallas, and Jacksonville, joined the lawsuit in June 2014 after suffering partial renal failure in April, despite no history of kidney problems. During one season with the Chargers, Wiley was diagnosed by team physician Dr. David Chao as suffering from a severe groin strain. He was treated with "multiple injections" of painkillers throughout the season to cope with the condition.

Following the season, according to Wiley, an independent doctor diagnosed a torn abdominal wall that required surgery.

"You can't walk into a doctor's office and say, 'Give me this, give me that, just to get through the day.' Somebody would shut the place down," Wiley told ESPN.com. "But that's what was going on in the NFL. It's easy to get mesmerized, I won't deny that. There's this 'play-through-the-pain, fall-on-the-sword' culture, and somebody in line ready to step up and take your place.

"The next question when people hear about this stuff is: 'Where's the personal responsibility?'" he continued. "Well, I'm not a medical doctor, but I did take the word of a medical doctor who took an oath to get me through not just one game, or one season, but a lifetime. Meanwhile, he's getting paid by how many bodies he gets out on the field."

Dr. Chao stepped down as the Chargers' team physician in June 2013.

Wiley spoke to the culture of the NFL in 2003, while playing for the Chargers and doing a weeklong, first-person series with the *Los Angeles Times.*

"Sometimes you need a fistful of Vioxx. Any anti-inflammatory. You need it to survive," he was quoted as saying. "Vioxx is a beast. I love Vioxx. I'm going to invest in that company when I retire. When you get a shot of painkiller, it's not like when they're taking blood at a blood bank. They've got to grind that needle in there. I mean grind it. When you've got an injury and it's acute, they've got to go get it. They can't just shoot you up on the surface. The guys laugh at me because I'm in tears when I have to get one. I bite on a towel and just try not to pass out. I hate shots. I've fainted on shots before, so anything like that kills me . . . After you get it, there's some numbing and there's a placebo effect. You know it's doing something so you feel better already. I probably don't want to know what they're shooting in there, but it's probably some kind of anti-inflammatory. It's like a fire extinguisher. Something's on fire inside of you, and they've got to get that extinguisher in there to put out the flames."

The pressure to play often is more internal than external. Players want to be there for their teammates and will try to play through anything, even concussions. It's one reason Junior was supportive of Goodell's push, beginning around 2010, to make the game safer by imposing stiffer fines and penalties — and in some cases suspensions — for hits to the head of defenseless receivers and quarterbacks.

Junior felt that players needed to be protected from themselves, a point that was driven home by his own career. He played in 268 games over 20 seasons, had more than 1,400 tackles, and administered and absorbed countless collisions. Yet he never was diagnosed with a concussion. That's not to say he never sustained one, because he did. Many of them.

"I can't even count how many," said Gina. "After games, particularly away games when I would watch them on TV, we would talk about the game and how he played. Sometimes I would ask him what happened on a certain play or why something went wrong and he wasn't in a certain spot. He would say, 'Oh, I had a concussion. I just had to shake it off.'"

That type of thinking and attitude was common before concussions became a national conversation around 2010. Prior to that, players, coaches, and team executives rarely used the term. Instead, they relied on euphemisms. It was common to hear things like, "He got his bell rung," or, "He got dinged."

Ironically, many players considered it a badge of honor to be concussed. Junior referenced it in an early 1990s *NFL Rocks: Extreme Football 2* video.

"When you put on a good hit, it hurts you too," he said. "This game is a matter of inflicting pain on the other person. It's an ego thing. If I can feel some dizziness [after a hit], I know that guy is feeling double of what I feel. So, yeah, the hitting that I put on somebody else is always going to be judged by how I feel going back to the bench."

It is only in the last five to 10 years that the public has begun to understand the connection between concussions and dementia, memory loss, and depression. The NFL tended to pooh-pooh, sidestep, or dismiss any suggestions of a link before then. Sometimes it went so far as to shoot down its own studies. For instance:

"In 2009, the NFL funded a University of Michigan study that showed that former players between 30–49 were 19 times more likely to have Alzheimer's and other mental disorders than men of the same age," ESPN.com wrote. "But the league disavowed the study, saying that it did not specifically study dementia and was based on unreliable phone surveys."

Nate Jackson played tight end in the NFL from 2003 to 2008, and after each season he would save his helmet. When talk about a possible link between repeated brain trauma and long-term health consequences grew from a murmur to a whisper to a shout, he began studying the helmets. To the rear bottom of one earhole was a small, clear sticker that featured tiny print. It read:

WARNING: NO HELMET CAN PREVENT SERIOUS HEAD OR NECK INJURIES A PLAYER MIGHT RECEIVE WHILE PARTICIPATING IN FOOTBALL. Do not use this helmet to butt, ram, or spear an opposing player. This is in violation of football rules and such use can result in severe head or neck injuries, paralysis or death to you and possible injury to your opponent. Contact in football may result in CONCUSSION-BRAIN INJURY which no helmet can prevent . . . Ignoring this warning may lead to another and more serious or fatal brain injury.

The scary part for Jackson was that no trainer, doctor, or coach had ever spelled out said dangers to him — or even notified him or his teammates about the sticker and its warning. In fact, the league and its TV partners glorified the violence and big hitters of the game, featuring them in commercials, video games, and highlight material.

That sort of behavior was at the heart of a 2011 class-action lawsuit in which more than 4,500 former players claimed that the NFL hid from them the dangers associated with concussions. The case was filed six months after Dave Duerson, a former safety with the Chicago Bears, put a gun to his chest in his Sunny Isles Beach, Florida, home and pulled the trigger—but not before leaving a note asking that his brain be examined by the Boston University School of Medicine.

BU scientists discovered that Duerson was suffering from chronic traumatic encephalopathy (CTE), a degenerative brain disease that's triggered by repetitive trauma to the brain. Each trauma produces a buildup of an abnormal protein called tau, which causes the brain degeneration associated with memory loss, confusion, impaired judgment, impulse control problems, aggression, depression, and, eventually, progressive dementia. Typically, the first symptoms present themselves years or even decades after the trauma occurs or an athlete stops playing and involve issues with judgment, reasoning, problem solving, impulse control, and aggression.

Duerson, who played for the Bears, Giants, and Cardinals over an 11-year career, was known to be suffering from depression and other neurological problems. His death took on greater significance 14 months later when Ray Easterling, a 62-year-old retired safety who had played for the Atlanta Falcons, shot himself to death at his home. Easterling not only was known to be suffering from depression, insomnia, and dementia but also was the initial plaintiff in the class-action suit against the league.

The suicides of these players, combined with mounting scientific research on the dangers of repeated brain trauma, put the league on the defensive. Over a four-year period, Commissioner Goodell made player safety during games a point of emphasis, penalizing, fining, and even suspending defensive players who hit opponents above the shoulders. He also moved the kickoff up five yards, to the 35-yard line, in hopes of decreasing the speed and violence of the collisions that occur on the play. And for the first time, the league placed independent neurologists on the sideline during games and created uniform protocols that had to be met before a player could return to the field after sustaining a concussion.

All of this was done with an eye on protecting the players, the league said, but cynics wondered if the real reason was to protect the golden goose. With revenues soaring past $10 billion annually, the last thing owners wanted was more litigation that could scare away fans and sponsors. The league was at a point where it could no longer deflect and deny the link between brain trauma and neurological health issues.

In fact, as part of its $765 million settlement agreement in the class-action suit, it submitted documents that said its players were likely to suffer chronic brain injury at a "significantly higher" rate than the general population and also to show neurocognitive impairment at a much younger age. The documents said retired players between 50 and 59 years old developed Alzheimer's disease and dementia at rates 14 to 23 times higher than the general population in the same age range. They also stated that the rates for players between 60 and 64 were as much as 35 times the rate of the general population.

More damning, the PBS show *Frontline* reported in September 2014 that the Department of Veterans Affairs' brain repository had found CTE in 76 of the 79 former players it examined, including Duerson.

Junior was experiencing personality changes as he aged. Normally jovial and good-natured, he would become angry and confrontational in the blink of an eye. In 2010, for instance, on Cinco de Mayo weekend, he chastised patrons who failed to stand when the national anthem was played before a sporting event being shown on the big screens. That weekend he also walked into the kitchen area of his restaurant and fired the entire cooking staff. He was upset that it was taking too long for the food to come out, and when someone seemed to chuckle after he went on a rant about it, he fired everyone on the spot, shut down the restaurant in the middle of the day, then walked off and had a shot of Jameson and a beer with his general manager, James Velasco.

Junior rehired the staff the next day, at the urging of Velasco, but that sort of erratic behavior was becoming more common. It often followed days or nights of heavy drinking, as happened when he got into an argument with his girlfriend at the time, Mary Nolan. They first met when he was playing in New England. He pursued her as if she were a quarterback on the run, the difference being that she made no attempt to evade him. Their relationship was combustible, though, in part because she was unwilling to accept the excesses in his life — the women, booze, and gambling.

On one particular occasion their argument got loud. It started in the sushi bar at his restaurant, where both of them had been drinking. At one point he ended up yanking her by the hair, either out of anger or because he had lost his balance. The atmosphere was so tense that someone called the police because potential violence seemed to hang in the air.

Junior was gone when law enforcement arrived. His close friends who were with him at the restaurant looked into getting Mary a hotel room to keep the two apart. It was not the first time the two had to be separated — nor would it be the last.

The next morning Junior met with Bette Hoffman, the director of his foundation and a person he affectionately called Mom. He was remorseful. "I don't know what happened," he said. "But I know this is not who I want to be."

It was an emotional conversation that ended with Junior agreeing to get help. Hoffman phoned the Betty Ford Center, which treats patients for drug and alcohol abuse. With his consent, she made arrangements for him to enroll in the program. Typical of Junior, however, he failed to show. He told Hoffman that he did not want to leave Mary alone.

Later that year Junior and Mary got into another argument. They had been out drinking that afternoon, and when they returned to the house she found one of his cell phones showing a message from another woman. She was livid. The argument is alleged to have become physically violent. At about five o'clock, Junior phoned Hoffman to come pick him up. When she arrived, Junior and Mary were outside his house. Hoffman stopped and told him to get in.

"He got in, and she came around the car and grabbed the door handle, saying: 'You can't go. You can't go,'" Hoffman said. "I said, 'Mary, you guys need to cool down. He's coming to my house.' She kept saying, 'You can't leave me. You can't leave me.'"

Hoffman drove Junior to her place, 20 minutes south in Del Mar. She made him a chicken-salad sandwich, and they watched the Sunday night football game between the Eagles and 49ers. During this time Junior's phone was blowing up with calls from Mary. Sometimes he answered, but often he hung up on her. So Mary began calling Hoffman's house number.

"'He's gone to bed. You should let things cool off,'" Hoffman said she told Mary. "She yelled profanities at me: 'You're a fucking bitch. Nobody can stand you. Everybody is nice to you because you know Junior, but he hates you. He's been trying to get rid of you for years.' I just said that I wasn't going to listen to it and hung up."

Junior heard most of the conversation because Hoffman had it on speaker phone. He knew there would be no peace unless he spoke to Mary (who declined to be interviewed for the book), so he called her back. It was roughly 10 o'clock. A policeman answered the phone and told him he needed to return to the house because a woman was filing a domestic violence complaint against him. Hoffman drove Junior back to Oceanside and was greeted by a line of police cars when they turned onto The Strand.

Oh shit, Hoffman thought to herself.

The police removed Junior from Hoffman's car and put him in handcuffs. They informed him that Mary Nolan was accusing him of striking her. When she spotted him outside, Mary began screaming, saying at one point: "Don't come back!"

Junior was taken to the Oceanside Police Department and booked on suspicion of felony spousal assault with injury. Hoffman posted bail and drove him to her house, then back to his, where he wanted to get some clothes and a car. They were supposed to meet at a coffee shop later that morning—she had made arrangements for him to stay at La Costa Resort and Spa—but it never happened because Junior drove his Cadillac Escalade off a 100-foot bluff in Carlsbad.

In the hospital, where he was treated for cuts, bruises, and a concussion, he assured family and close friends of two things: he did not strike Mary, and he did not attempt to kill himself. He said he had fallen asleep behind the wheel, which police said was consistent with the lack of skid marks where his car went over the bluff.

Family members wanted to believe him. Junior was known to get sleepy behind the wheel. In high school he had totaled a truck when he fell asleep and drove off the side of the road. His mother still shook her head about it. The top of the vehicle was nearly severed off that day, and when she saw it she thought to herself that there was no way anyone should have survived.

"God had His hand on my son," she said.

Melissa Waldrop, Junior's girlfriend in high school and college, often did the driving when they returned home from Los Angeles because he would get sleepy behind the wheel.

But on this day Junior's inner circle knew there would be a media frenzy and questions about whether he had tried to commit suicide. It was decided that he couldn't return home because reporters would be waiting. He would have to stay with one of them, but which one? Sadly — and significantly — Gina was the only person in his inner circle of friends who offered to take him in.

He spent several days in her expansive hilltop home, sleeping in a guest room that had the same furniture the two had shared in their bedroom while married. Gradually and softly, Gina tried to walk him through what had happened. His story was not adding up.

Finally, after two or three days, she told him they needed to get out of the house and get some fresh air. She suggested they take a drive. He was against it initially, but eventually agreed. He put on a hooded sweatshirt and pulled the top over his baseball cap. His eyes were covered by dark sunglasses, and he slouched low in the seat.

As they drove, Gina gently asked him once more to recount what happened the night he went over the bluff. When he didn't speak, she would recite out loud what he had told her. Eventually they found themselves on the street approaching the bluff. Junior was visibly tense and uncomfortable.

"Get me out of here!" he said. "Hurry up!"

More than ever, Gina was having trouble accepting his story. That night, when things had calmed, she asked him about it again. It was two or three in the morning. Their three kids were sleeping. The mood was serious yet nonconfrontational. It was just them and the truth.

"Was there any part of you that wanted to drive off that cliff? Did you do that on purpose?" she said. "Tell me the truth."

There was a pause, then an empty, glazed expression fell over Junior's face. During the silence Gina's mind raced with what she knew: *He's a terrible driver. He falls asleep in the car all the time. It was late at night. He was tired.*

Then, for one terrible, frightening moment, it hit her: *He did do it on purpose.*

"You can't do that!" she said. "Don't ever do that! You have three kids. We love you. We need you."

Gina began crying.

"I'm so concerned about you," she continued. "I want you to never feel you have to drive off a cliff again. What can I do to help you? I'll drop everything right now to make you feel and know that you *are* loved."

With sad eyes, Junior shifted his jaw, looked away, and said: "I don't think I'll ever know that feeling."

Gina wasn't the only one suspicious of Junior's story. Various family members also confronted him about the incident.

"I saw him in the hospital, and he said, 'Sis, I'm sorry. Don't believe everything you hear,'" Annette said. "I thought he was groggy and told him to lay down and don't worry about it. But a few days later I told him, 'Okay, now I'm going to ask you, because people are talking. Was it or was it not [a suicide attempt]?' And he said, 'Sis, I told you not to believe everything you hear.' He goes, 'I was tired. I was braking.' 'So, okay, it wasn't? I need to know.' He said, 'No.'"

Annette accepted his word, but after a week of listening to public speculation that he had tried to take his life, she confronted him again.

"I can't ignore it," she said to him. "I'm going to ask you again, did you?"

"Sis, I wouldn't do that to Mom and Dad," he said. "I wouldn't even do that to you guys. I didn't do it, and I wouldn't have thought to do it."

"Okay, Bug," she said. "But you really need to see someone. I told you that you were not going to do well retiring. I told you that you were not going to transition well. I asked you before, when you retired, and I'm going to ask you now: Please go get help. Please go see somebody."

Annette explained later that "there were times when he would blank out. We'll be sitting there and he'll just totally blank out. When he snaps out, there's like a depression look on his face. [I'd say], 'Bug, are you depressed?' He'd say, 'No, no. I'm fine. I just need my family.'"

Annette cautioned him that family wouldn't be able to fill that void all the time.

"I honestly think you need to see somebody, whether it be a doctor for depression or somebody to help you transition over because you're not in that limelight," she said.

"I don't need that limelight," he responded.

"I honestly think you need help," she answered.

The last time they spoke about it was Thanksgiving of 2010.

"Sis, stop," he said. "Stop being a mother."

"I'm not being a mother," she said. "I'm just being a worried sister."

Junior protested that everything was fine.

"It's not," she countered. "If you're calling me late in the evening or early in the morning to come to your house, you can't tell me you're okay. I love the fact that you're spending time with family, but there are times where you are as blank as day."

Annette thought about the conversation months later. "He could be around family and enjoying himself, then all of a sudden, with the snap of a finger, he's got that blank face," she said. "Then, with another snap of the finger, he's out of it."

Rev. Benson Mauga, one of Junior's uncles, also had concerns about whether Junior had tried to take his life. Mauga was out of town when the incident occurred, and when he returned he wanted to get Junior alone with his parents to address the rumors. Unannounced, Mauga drove Mama and Papa Seau to the restaurant to meet with Junior, who was in his private office.

"I want you to come sit down," Mauga said to his nephew. "It's very serious what I have to ask you. You're probably tired of hearing it. I want you to say it right in front of Mom and Dad. Did you really do it?"

Junior was silent for a moment, but those couple of seconds were an eternity for Mauga.

"From there, I knew that he probably had some kind of thought about it but didn't want to confess it in front of them," Mauga said. "He said, 'No, I fell asleep.' But I already saw it in his eyes. He just didn't want to hurt the parents.

"I said to myself, *Okay. I see it. I feel it,*" Mauga continued. "I could sense it even by the way he got up and stepped from around the desk. He did it slowly. It was like he's thinking, should he admit it right there when he had the opportunity with Mom and Dad — while I was there to comfort and slow the dad? His dad is very quick with anger, and it might erupt in the wrong manner. But he didn't do it. He told them he fell asleep."

Privately, Junior told two people who didn't want to be quoted in the book that he did indeed drive off the bluff intentionally. Either way, the incident was a clear sign he was in need of help. But immediately afterward he was unwilling or unprepared to accept it.

"We talked a lot those several days that he stayed with us," Gina said, "and I was saying to him, 'This is a chance, you have another chance. You're lucky to be alive, you're lucky you're not in a wheelchair or something horrible didn't happen to you. You can walk away from this incident, and God's given you another chance. Let's make it count.' I was trying to be positive and help him see that there's a lot to live for, and there's a lot of good things happening around him . . . and there's so much possibility out there for him. I was met with a really blank stare. I know he was exhausted and tired from the accident, but there was just blank. It's like he was looking at me but not really connecting."

One person who did manage to connect with Junior was Aaron Taylor, his former teammate. Taylor had concerns following the accident, so he reached out in a supportive yet non-intrusive manner.

"I sent him a text: 'Hey, man. Heard what happened. You alright? Letting you know if you want somebody to talk to, or just listen, I'm here,'" Taylor said. "He reached out to me by phone. He said, 'I need help. I'm an addict. I can't stop drinking. My life is spinning. I don't know what's going on. The pressures are getting to me. I don't like the man that I see when I look in the mirror. I've done fucked up shit as a father.'"

Junior told Taylor how he had promised to take son Jake to a lacrosse tournament, but failed to do so because he was hungover.

"He had a lot of shame and a lot of guilt about that," Taylor said. "He cried when he told me that story. As a dad with sons, I remember it being powerful. That was as low as I had ever seen him. I remember playing against Junior, I remember playing with him — he was the iconic alpha male warrior, the toughest of the tough. But he was as broken as I've ever seen a person, let alone him, which in my business of recovery means there's hope. I was like, *Sweet! This cat is ready.*"

Taylor suggested that they attend a 12-step meeting. Junior agreed. The session was at a local church, with eight to 10 people present. Taylor was nervous about whether Junior would feel comfortable enough to open up. It's one thing to be John Doe admitting your frailties and vulnerabilities to strangers. It's quite another to be someone whose name and face are as synonymous with San Diego as Sea World or the San Diego Zoo.

Hi, my name is Junior and I'm an addict, an alcoholic. I'm out of control. I've got a lot of shit going on.

"He shared feeling a lot of pressure from his family, everybody always wanting something from him, letting people down around him," Taylor recalled. "He didn't get right down to the nitty-gritty, but I was shocked at how open and candid he was about how he was feeling, the sense of not feeling in control and what seemed, to me, to be an awareness and an embracing of where he was at."

They attended another meeting at a sponsor's garage. About 12 to 14 people were in attendance. Again, Junior opened up and spoke about his addictions and the pressures he was feeling. He also was attending Bible study classes with other friends. Taylor felt a sense of pride because he thought his friend was on the road to recovery. But typical of their relationship, Junior went off the grid and stopped responding to his messages. He also would go for long stretches without speaking with Gina or the kids.

"We heard from him generally every week, sometimes several times a week," Gina said. "But then he'd go several weeks without any communication. The longest span he ever went without us communicating was about two and a half to three months, and I asked him, 'Where have you been? What's been going on?' And he just said, 'G, I'm just in a really dark place.' His exact words I remember were, 'I'm so dark that even picking up my surfboard doesn't make me happy. That wouldn't even put a smile on my face.'

"That was horribly concerning because the one place of solace for him was to get in the ocean, and he lived right on the beach there in Oceanside. That was his one form of peace, of gaining just a moment of quiet time. It always made him happy because all the guys were out there surfing, and he always went to breakfast with them. He always had a place that he could go where there were buddies around, and camaraderie with friends and other surfers, and going to the gym. And he just said he didn't want to do any of that."

Junior's financial decisions were also becoming a concern. After being defrauded by Gillette in 1996, he turned to local financial planner Dale Yahnke to oversee his portfolio. The two bonded in part because Junior wasn't looking for the big splash. He was typically conservative with his money and his investments. He and Yahnke hit it off because, like Junior, Yahnke liked to think long-term.

Yahnke's goal was to make sure of one thing: that when Junior retired, he would have to work only if he wanted to, not because he needed to. But the young man who had always been so mindful of his money, who used to have nightly receipts from the restaurant faxed to his dorm room during training camp so he could review the numbers at one or two in the morning, was becoming increasingly erratic with his spending.

He was taking more trips to Las Vegas and San Diego County casinos, which coincided with his failed investments in a project to build 15 Ruby Tuesday restaurants in Southern California over a five-year period. His advisers, friends, and family all pleaded with him not to do the deal, saying the risk was too great. But he proceeded anyway and lost big.

He initially invested $300,000, but only two of the restaurants were built, and each was a money pit. Compounding matters was that he had made guarantees on the leases that kept him on the hook even after the businesses failed. Yet despite those disasters, he still could have landed on firm financial ground if not for his regular trips to casinos.

The Bellagio and Caesars Palace in Las Vegas were two of his favorite spots. Some family members believe that he gambled in an attempt to recover losses from the restaurants, but others who spent time with him at the tables said he loved the adrenaline rush of chasing the next big win.

Casinos attract highly paid athletes by presenting the facade that the players are getting something for nothing. They send private jets and put them up in presidential suites, all at "no charge." Then they give the athletes a line of credit based on their income or worth—for someone like Junior it could be millions of dollars—with the understanding that they wouldn't have to repay any losses (known as markers) until the casino requested repayment.

Junior was known to sit alone at a blackjack table in the high-rollers room and play five to seven hands at a time, often with chips valued at $5,000 or higher. His personal driver, known as Big Tony, claimed to have seen Junior go up $2.7 million in 20 minutes, then lose more than $3 million in the next half-hour.

U-T San Diego reported that over his gambling lifetime with MGM International, which owned the Bellagio and Caesars Palace, Junior had a net buy-in of $6.8 million for all of its properties—$4.1 million of which came from markers. By the end of November 2010, *U-T San Diego* reported, he owed $800,000 in markers to Caesars Palace and $500,000 in markers to the Bellagio. His average bet at the Bellagio was nearly $39,000.

"I knew that what he was doing in Vegas was going to end only one way, and I thought it would be humiliating for him," Yahnke said. "I tried to talk him out of it. He didn't listen."

"We landed in Vegas one time, and immediately, within hours, he won eight hundred something thousand dollars, okay?" said Junior's friend Jay Michael Auwae. "So he comes back up to the room, and I said, 'Let's go home, surf, chill, pay some bills.' But after dinner a whale-watcher [a casino handler charged with roping in big-money gamblers] comes up to the room. I'm saying, 'June, enough already.' And he goes, 'No, bro. One more time. I'm gonna clip 'em.' Not even two hours later, he comes back up and hits the table with a glass and starts cussing. I was like, 'Please don't tell me . . .' He had lost it all. He's lying on his bed looking at the ceiling, and I go, 'Buddy, you gotta stop this, man.' He goes, 'We got this. We'll get 'em tomorrow.' The next morning the whale-watchers show up. June got another half-million dollars [up], and he goes back down and loses the whole thing."

16

High Point, Low Moment

ON AUGUST 27, 2011, the Chargers announced plans to make Junior the 35th inductee in their Hall of Fame. It was their way of paying homage to not only one of the game's all-time greats but also a community icon. The announcement was accompanied by a written statement from team chairman and president Dean Spanos.

"From the day we drafted Junior, we knew he was special," it read. "He's such an energized, charismatic person. He attacks life the same way he attacked ball-carriers. It's that passion that turned him into the Hall of Fame player he is. His athletic ability speaks for itself, but it's his passion and energy that separate him from the rest of the league. It's that same passion for life that has made him an icon in San Diego. He *is* San Diego in so many ways . . . in how he has represented himself on and off the field, and how he has been there to help and support the community when and where it's needed. Junior is such a unique person, so full of life, and that rubs off on everyone around him, including me. I'm blessed by his friendship."

The ceremony would take place exactly three months to the day later, at halftime of the team's sold-out game against the visiting Denver Broncos. Typical of Junior, he acknowledged the honor with humility and humor, thanking the Spanos family but adding: "I think Dean still owes me a dollar from the golf course, but we won't bring that up."

"You don't plan on something like this," Junior said, in all seriousness, a few days before the ceremony. "You just go through your journey and hope that someday you get a little love back and some respect. But the honor that we're receiving Sunday is going to be overwhelming. I don't know how I'm going to react. I really don't."

When November 27 arrived, a blanket of sunshine fell over the city. It was a perfect afternoon for a game and a tribute. Junior planned to meet a few family members and friends at his restaurant beforehand, then take a limo to Qualcomm Stadium. He was dressed in a dark-blue suit with faint pinstripes, a white dress shirt, and a patterned lavender tie. He also wore a Samoan money lei around his neck.

It was supposed to be one of the greatest days in his life. It turned out to be one of the worst.

It began with his son Jake refusing to accept an invitation to attend the event. Jake was still angry that his father had failed to reciprocate with the same kind of support and compassion for Gina that she had shown for him after the cliff incident. The backstory: Gina had suffered a serious hamstring injury that required surgery, followed by three months on crutches. She asked Junior to take Jake, the second of their three children — Sydney was the oldest, Hunter the youngest — to a lacrosse tournament while she was rehabbing. He agreed, but failed to show up the morning they were scheduled to leave.

Jake was livid. Another family member took him to the tourney, and the next time he saw his father he confronted him. Jake's anger was not about the one incident, but an accumulation of occasions when Junior had failed to spend time with the kids.

"It was a combination of things, and me growing up and becoming aware of our relationship — or lack thereof," Jake said. "It was a moment where I decided I was sick of waiting and I was going to bring it up to him, and I wasn't going to let him do what he normally did, which was kind of ignore what you were saying. That was a big step in our relationship, me trying to reach out to him and not give him the option of pushing it under the rug.

"He didn't take it very well, but I wanted him to know that we wanted him around more. I thought we could see him more often and spend more time with him than we were getting. He didn't like that I called him out on it. I probably didn't say it very gracefully either; I was pretty angry and pretty aggressive. I doubt that a lot of people had spoken to him like that in a long time."

By the time the Ring of Honor ceremony rolled around, his emotions from that argument remained raw, which was why Jake decided to stay home.

"I don't think it would've been appropriate for me to go at that time," Jake said. "I wanted to support him, but at the same time I wanted him to know that what I wanted from our relationship was more than he was giving. That was a way for me to do that."

Tension between Junior and his kids was nothing new. In their eyes, he often allowed football and socializing to take priority over them. He would promise to do something with them, such as attending one of their athletic events, then arrive late or not at all. Even in serious situations he could not be counted on to be there, such as in September 2010 when Sydney had an appendicitis attack and was taken to the hospital. Gina phoned Junior so he could join them, but he was drunk at the Taste of La Jolla and failed to show.

The strained relationship between him and Tyler — the son he had with his high school and college sweetheart Melissa Waldrop — was the most volatile. Tyler rightfully felt like the forgotten one — or at a minimum, the overlooked one. Media reports often talked about Junior's three kids, ignoring Tyler's very existence. If that was not painful enough, there was the fact that Junior, early in his marriage to Gina, had sought a paternity test to confirm that Tyler was his son.

Everyone knew the child was his. He and Melissa had met as teenagers, and she was crazy about him because he was her first love. Teammates at USC used to smile at how he would carry Tyler in one hand, tucking him against his body as if he were a football.

Only Junior knew why he requested the test, but he later told Melissa that he regretted doing it. It caused a strain not only between the two of them but also between him and Tyler when Tyler got older. Tyler often felt as if he was not as important to his father as his other kids; Junior did not regularly visit him on holidays, and when he took his family to the Pro Bowl, Tyler was never included. Junior shrugged it off by saying he didn't want the "drama" of dealing with Melissa.

In March 2010, Tyler confronted his father about the years of "living in the shadows" of Junior's family. "I told him, 'I want you to be part of my life. If you want to be, that's awesome. If you don't want to be, it's okay, but I need to know. I don't want to wait around to hear from you, and be mad and upset.' He told me, 'We'll never get those early years back.' I cried, and he cried. But I didn't feel that emotional connection with him."

Instead of improving, their relationship continued to fracture. Junior was upset that Tyler had gotten his girlfriend pregnant while at Delta State College in Cleveland, Mississippi. He had spoken to him about not following in his footsteps of being an unwed parent. When Tyler left school before graduating, Junior was furious. He gave him a job at the restaurant, but he rode him hard and showed no favoritism.

Sydney? She was another story.

"The tone of his voice changed when he was talking to Sydney," said former girlfriend Megan Noderer. "He always ended calls with her by saying, 'You know I love you, right?' That was a line reserved for Sydney."

She had his heart not only because she was his only daughter, but also because she was a virtual replica of him. Her smile could fill a room, just as his could. And her spirit was as bright as her eyes, just as his was. She also possessed that special ability to disarm people with her words and gentle touch, just as her father could.

The two were so close that Sydney often made excuses for Junior when he failed to call or visit. But in late 2011, after not hearing from him for nearly four months, she too confronted him.

"I was used to him taking breaks and then randomly texting me and telling me that he loved me," she said, "and from there I'd fall right back into it and forgive and forget and do the whole thing over again. But this time was different. He literally fell off the face of the earth and didn't text me, didn't call me, didn't try to reach out to me for four months. It freaked me out."

In the meantime, Sydney had to write a parent profile for an advanced-placement language class in school. She chose to focus on Junior and listed all her resentments toward him and frustrations with him. When she finally stopped writing, she had filled 11 pages. Later she used the contents of the profile to write a letter to her father; then she set up a meeting with him so she could deliver it personally.

"I just texted him and told him that I would meet him at his house at a particular time, and I did," she said.

She hadn't planned to say much before handing him the letter, but almost immediately she began speaking from the heart.

"If I'm the one girl in your life, why can't you make time for me?" she said. "All I do is try to be the adult in the situation and find time for you and love you and show you how much I need you in my life, and you just decide to ignore me, and it hurts because all I want to do is make you happy."

Junior, with tears streaming down his face, sat on a couch and said nothing.

"He just sat there and looked past me at the wall," Sydney said. "He didn't come to console me, didn't touch me, didn't ask if I was okay. I just spilled everything for 30 to 40 minutes and then asked him what he thought. He looked at me and said, 'Syd, I don't know how to love.'"

Sydney was shocked.

"'What do you mean?'" she said she told him. "'I love you so much, and I know you love me too. Why is it so hard?' And he was like, 'I don't know how to do it anymore.' He was breaking down. That's when I knew something was wrong, because he was shutting down as a person. I could tell he wanted to reach out, and he wanted to be the person he was before, but he couldn't. Physically and emotionally it was becoming too much for him."

When there were no more words to say, despite the two failing to come to some sort of resolution, Junior told Sydney he wanted to take her to dinner at a hole-in-the-wall Mexican restaurant so she could try the gumbo.

"Gumbo? I'm like, 'Are you serious?'" Sydney said. "Of all things, I just finish crying my eyes out, they're puffy and red, and he wants to take me for gumbo. I barely touched any of it. Then he made sure I got home safe. Every time I got in the car he had me text him immediately when I got home, and if I didn't text him he'd call and say, 'Okay, you know I love you, right?' And I would say, 'I love you too. I'm fine. Go to bed.'"

Sydney was always the quickest to forgive her dad, and as the Ring of Honor ceremony approached she sought to have all of her brothers attend. Jake continued to refuse the invitation, though, so only Tyler and Hunter agreed to go in the end. Everyone was supposed to meet at Seau's The Restaurant and ride to the stadium in the limo, but as was typical of big events in Junior's life, more people showed up than were invited — and *everyone* wanted a spot in the limo.

Before long, attitudes flared and tensions escalated. Frustrated and angry, Junior put his family in the limo and had a staff member call him a cab, which he took to the stadium. On the ride over, he phoned Gina.

"Here I am, throwing my own party again — free drinks, free food," he said. "Are you sure you and Jake aren't going to come?"

"No, I'm going to support Jake," she said. "Are you okay?"

"I can't wait to get this fucking day over with," he said.

No one at the game knew what was going on, but there was an air of uneasiness surrounding Junior. Normally charismatic and preacher-like when holding a microphone, he went through the motions during a speech that lasted just over three minutes. He thanked ownership, the equipment staff, and the team doctors. He prayed for the servicemen and -women around the world. He asked kids to work for today, build for tomorrow, and pray for the rest, the familiar tagline he attached to his public speeches.

But the passion was not there. The speech was more like painting by numbers. It lacked passion and purpose.

"I had no idea what strain he was under because he wouldn't tell me anything," Sydney said. "Of course, I could tell something was going on by the way he reacted to people. He wasn't, like, alive. He was just going through the motions."

When the ceremony ended, Junior walked off the field, down a dim tunnel, and out to the parking lot. He was accompanied by Megan Noderer, his girlfriend of only two months, Megan's sister-in-law, and his son Hunter. Unable to locate his limo, he loaded everyone into a cab and headed back to the restaurant. The place was bustling with activity, but he wanted no part of it. He stood along the second-floor railing outside his office and gazed at the people below.

"I am the inductee, I am the one who people are supposed to be celebrating, and everybody is mad at me," he told James Velasco, the restaurant's GM. "The city of San Diego is celebrating me, and I have to throw my own party."

It wasn't the first time his heart had been filled with equal parts frustration, sadness, and anger because he felt unappreciated by family. Melissa Waldrop, the girlfriend who knew him before the rest of the world did, remembers being with him in the VIP area during a party at his restaurant.

"He looks at me and says, 'I don't know a single person in here,'" she recalled. "I said, 'Yeah, but you know me.' He looked at me and said, 'You're the only one that really knows me.' Then he embraced me and hugged me. There's always been that burden since early in his career that he felt a responsibility to take care of his family. It was like a ripple effect. It went beyond just taking care of his parents and kids. It was everyone."

On one of the biggest days of his life, the man with so much life and passion watched, unsmiling, as other people celebrated. Then he slipped out of his restaurant with Noderer and her brother and his wife, got in a car, and drove home. He may have been surrounded by many, but he felt alone.

A week or so later, he and Jake met for breakfast and talked things out. "He apologized and said that I was right," Jake recalled. "My motive was to spend more time with him, and the way things were, I was spending absolutely no time with him. It was interesting. We moved past it, but I don't think everything was necessarily *fixed*. We settled the fight, but nothing really changed."

Junior and Jake had no contact over the next few weeks, but on Christmas Day 2011 Sydney persuaded Jake to go with her to Junior's house. Jake didn't want to go; he was still dealing with his anger and frustration. Junior beamed when the kids arrived. For the rest of the evening they all laughed and hung out with friends who were visiting and had a good time, so much so that Sydney and Jake spent the night.

"We were dancing and singing and listening to music," Jake said. "It's one of my best memories with him. It was probably the first time we hung out together where we weren't at a formal dinner or something. We just had fun."

17

"I Knew He Was Going to Have a Hard Time with Life After Football"

DURING ONE WEEKEND in April 2012, Hunter Seau spent the weekend at his dad's beachfront home. Junior was making an effort to be more accessible to his kids, and Hunter, a thoughtful and sensitive 11-year-old, loved when they could just hang out and enjoy the beach or visit the golf course.

On this particular night, when Hunter got out of bed at 3:00 AM to let the dog outside, he noticed a light in his dad's room. When he peeked inside, he saw Junior sitting on the edge of the bed, staring at a TV that was not even on.

"Dad, are you okay?"

"Yes, son. I'm fine."

Junior had suffered from insomnia dating back to high school. His condition was particularly acute during his NFL career, when he had trouble shutting off his mind the night before games. He often coped by taking Ambien, the most commonly known brand name for the prescription drug zolpidem.

Zolpidem is not without potentially serious side effects, though. The manufacturer cautions that suicidal thoughts or actions have occurred among people who have taken it, and the accompanying instructions warn that it should not be taken with alcohol.

Junior drank regularly—and sometimes heavily. His alcohol of choice was Belvedere vodka, but later in his life he also took a liking to Jameson Irish Whiskey. When he couldn't sleep after an all-night bender, he sometimes would pop a sleeping pill to help him rest. The danger: depression is one of the potential side effects when mixing Ambien and alcohol.

Mary noticed a difference in her brother earlier in the year and was concerned about him.

"You look tired," she told him. "What's going on? What's wrong?"

"Nothing," he said.

"Something is going on," she said.

"You know me too well," he answered.

The two talked about the family and financial strains he was feeling. It was wearing on Junior, who for years had been stressing to his parents and close family members that they needed to stop looking at him as the highest-paid defensive player in the NFL. He was now retired, with a significantly reduced cash flow.

Annette, his other sister, began realizing something was wrong even before Mary did. The most telling sign came in August 2011, when he allowed her to pick up the check after she and her husband had dined with him and a female friend in Dana Point, California.

"Junior never allows me to buy anything, so when we went out this time I said, 'Put your wallet away, I'm paying. I'm tired of you paying for everything,'" she said. "When he put his wallet away, my husband said, 'Something is going on.' It was weird that he let me pay."

Annette did not push it, but three months later she began to put the pieces together during another conversation with him. She had told him her family was thinking about moving out of the house they shared with Mama and Papa Seau because they needed more space. Junior gently pleaded with her to stay.

"You can downsize and buy Mom and Dad a smaller place," Annette said.

"I need to check my finances," he responded.

When Annette joked that Junior wanted her to stay because he lacked the money to purchase another place, he shot her a stunned expression. It was as if a deep secret had been exposed, and Annette immediately was concerned.

"Sit your butt down. You and I are going to talk, and you're going to stop trying to be this prideful man," she said. "You're going to file for bankruptcy, aren't you?"

"No, I'm not," he said.

"Ahhhhh . . . okay . . . I got it now," Annette said.

"What? What do you got?" Junior demanded. "Who's talking to you?"

"You're going gambling to try to win all this money that you're losing on the restaurants," she said. "You're killing yourself because I know your restaurant is hurting. You hardly have people in there, and you have another [competing restaurant] that opened next to it. If you're dry, file bankruptcy before creditors take you to town."

"I don't know what to do," he said. "I just don't want Mom and Dad to get mad. And my kids—I don't want them to think bad of me."

Annette reminded him that the worth of a man is not measured by the size of his wallet.

"We grew up without money," she said.

"But I promised Mom and Dad that they were going to have a better life," he said. "I just have to make sure that Mom and Dad are taken care of."

Months passed without Junior doing anything to address the situation. He was slipping into a deeper and darker place as his annual golf tournament approached in March 2012. The gambling, the drinking, and the collisions from a 20-year career were taking their toll on his body and his mind, as reflected in a comment he made about the perception that Commissioner Roger Goodell was making the game "softer" by instituting enhanced player safety rules.

"It has to happen," Junior said. "Those who are saying the game is changing for the worse, well, they don't have a father who can't remember his name because of the game. I'm pretty sure if everybody had to wake with their dad not knowing his name, not knowing his kids' names, not being able to function at a normal rate after football, they would understand that the game needs to change. If it doesn't, there are going to be more players, more great players, being affected by the things that we know of and aren't changing. That's not right."

The changes in Junior were becoming more obvious, even if those around him refused to acknowledge them—or chose not to challenge him on them. One exception was former Chargers coach June Jones, who witnessed the excesses in March 2012 after arriving in town for the golf tournament.

"I knew he was a guy who was going to have a hard time with life after football if he wasn't involved with the game, and I was always trying to get him to come coach on my staff, but he would put me off," Jones said. "This time we had a long, long conversation about a lot of things. I called him on the carpet for some of the alcohol and some of the things I was witnessing that I knew were not him. For the first time, he kind of put his head down and wouldn't look at me. He had always looked me in the eyes before. He said, 'I'm all right, brother. I'm all right. I've got it under control. You can count on me.'"

Jones again reached out and extended to Junior a spot on his coaching staff.

"He looked me in the eyes and said, 'June, I promise. When you ask me this next year, I will come. I will do it. I got some things I've got to take care of first,'" Jones said.

That same night, at the annual banquet dinner, a somber Junior did something he had never done before: he started the event by calling his entire family to the stage to perform a Samoan hymn.

"He always started off by recognizing his mom and dad, but this time he brought everyone up on the stage," said Mike Norris, his long-time personal photographer. "He said, 'If you're here, come on up. Let's do this.' And they did a big prayer together. It was a very, very touching moment. There was something deep about it."

A month later, the Bellagio casino called in a $400,000 marker that was returned due to insufficient funds in his personal account. Junior's financial world was collapsing around him, but he refused to let anyone know. On Sunday, April 29, he went to his local workout spot, The Gym, for a memorial service for Marty Miller, a postal carrier and gym patron who had died of cancer. Junior, long known for his inspirational messages, gave the eulogy and talked about loving life and making a difference.

Afterward, he went to lunch with Mark Walczak, a former teammate who was visiting from Scottsdale, Arizona. Junior had forgotten his wallet and called Noderer, his girlfriend, to see if she could bring it to him.

Noderer was unlike any girlfriend he had had since the divorce. One, she wasn't blond. Two, she had a college degree and a career as a sales executive for an auto company. And three, she was not in her early twenties; Noderer was actually in her thirties.

She and Junior met in October 2011, at the wedding of Noderer's brother, Jason Zitter, who went by the nickname JZ. Junior and JZ had been friends for years. They hit it off at a charity golf tournament and started hanging out afterward. For a long time Junior didn't know that JZ had a sister, and when he found out he was immediately told by JZ that Megan was off-limits.

But Junior was one of those guys who hated to be told he couldn't do something. It was like putting candy in front of a child and telling him not to touch it. So he kept pushing JZ to introduce him to Megan. When JZ would be on the phone with her, Junior would playfully plead to speak to her.

At this point it wasn't so much that he was interested in dating Megan, whom he had never met, but about the joy he took in seeing JZ squirm. "Aw, come on," he would say. "Just let me talk to her."

Megan was not a big sports fan and knew virtually nothing about Junior beyond what she heard from her brother. She used Google to fill in the blanks. When the two finally met several days before the wedding in the Florida Keys, there was instant chemistry.

"The first night we started talking, I was like, 'I'm not even allowed to talk with you, let alone be alone with you,'" she said to him, laughing. "When we figured that we might kind of like each other, we were both so nervous. I said, 'You've got to ask my brother, because I'm not telling him.' He was like, 'I have to tell him?'"

On the morning of the wedding, Junior phoned JZ and asked him to go on a quick run. At first JZ passed, but Junior insisted he had something to talk to him about.

"Do not tell me you want to date my sister," JZ said.

"I promise you that I won't even try to kiss her while we're in this area code, but I think I like her," Junior said.

JZ shook his head and said, "You're killing me."

Megan was instantly intrigued with Junior because of his smile. It was so real and infectious. But she began to fall for him after seeing how far he went to ensure everything was perfect at the wedding. He had gotten ordained so he could preside over the service, and in the lead-up to the wedding he focused on the smallest details, like trying to learn everything he could about JZ so he could incorporate it into the service.

"My brother and I had lost our parents when we were young, and leading up to the wedding, Junior wanted to know so much about them to bring them into the service to make it special," Megan said. "It was overwhelmingly touching. He cared so much about my brother and wanted to make it right for my sister-in-law."

The one thing Junior could not control was the weather, which was miserable. Rain was blowing sideways the day before the wedding and still coming down the morning of the day itself. It was so bad that Junior and Megan spent the morning of the service rearranging chairs and speakers three different times, all but firing the wedding planner.

"He wasn't stopping short of anything; he was going to make it special," she said of Junior. "One of the biggest things I learned about him was, you want him on your side. I've always been a very career-oriented, focused person, and he used to joke when he introduced me to people — 'This is "Red," and she has a *career*. And she's over *thirty*.' I wanted him on my side because he was my biggest cheerleader.

"I remember nights where I'd travel and I'd be up late at night rehearsing for a pitch, and he'd call me and coach me through some things. Even my boss, I'd put him on speaker and Junior would give us a pep talk. He would set his alarm, no matter what time zone I was in, to make sure he called beforehand if I had an important meeting. It could be four in the morning and it didn't matter. He was my cheerleader."

Megan lived in Dallas, and she and Junior made a pact that they would never go more than 10 days without seeing each other. When that was too long, he would call and ask if she could come early. One such occasion was the final week of April 2012.

"He said, 'Can you just hurry up?'" Megan remembered. "I'm like, 'Why? I've got work to do. I'm busy.' And he said, 'Because when you're here everyone else leaves me alone.' It was a constant that everyone thought they were his friend. It was exhausting for him, just flat-out exhausting."

On this occasion Junior wanted a break after spending a couple of days with Walczak, but he was too nice to say it. Instead of being honest, he did as he typically did and smiled and acted as if everything was fine. He knew Walczak was scheduled to leave that Sunday night, so he put on his mask and played the role of good host.

He took Walczak to lunch, and from there, the two traveled to Hunter Steakhouse in Oceanside, where they hung out until Walczak left by himself around 9:00 PM. Walczak thought he'd see Junior before leaving town in a couple of hours, but Junior didn't return until much later.

The next morning, with no one at the house but Megan and himself, Junior spent the early morning in the ocean just beyond his front door. The open water represented peace and freedom for him. It was where he got away from everything — financial problems, women issues, family matters. There was only him and the waves.

"He would tell me that the only time he truly felt at peace was when he was with his children or in the surf," former teammate Rodney Harrison said. "He would say, 'When I'm on those waves, it's the greatest feeling. I have no worries, no stress, no problems. I just forget about everything.'"

Later that morning he and Megan made the short drive north to San Clemente, where he was scheduled to participate in a charity golf tournament and she was supposed to meet with a client in Orange County. Junior was Junior to those who saw him at the tournament — happy, effervescent, personable, relaxed.

"He would go up to everybody who worked there and say, 'Let's do a photo,'" said Mitchell Sacharoff, his cart-mate for five hours. "So rather than them having to ask him, he'd just say, 'Come on over here.' He was constantly in a great mood and taking care of everybody."

But this time it worked to his detriment.

"When I swung back by the golf tournament to pick him up and finish watching him play, he got in the car and I could tell he wasn't feeling well," Megan said. "I asked him what happened, and he said he golfed with this group of guys that was great and normal. He always said that about someone he liked, that they were 'normal.' Then he said he didn't feel well because the team wanted to do a dip and he couldn't say no."

By "doing a dip" Junior meant inserting a wad of chewing tobacco in their lower lips. Junior was not a tobacco guy, but he did it anyway.

"We were a team," he said to Megan. "But I'm going to be sick all night, and I just ruined date night."

"I was like, 'You're kidding me,'" Megan recalled. "But that was Junior. He did everything for others and not for himself. He knew it was going to cost him later—and that I probably wasn't going to be too happy. But he did it because it was what everybody else did and wanted him to do."

During the short drive home the conversation turned from sickness to sadness. As Megan steered the car down the highway, Junior steered the conversation toward his future and why he stopped pursuing the TV career he always envisioned himself having when he left the NFL.

"He told me he didn't feel he would ever recover his public image from that domestic violence allegation," Megan remembered. "He said that ruined it. People saw him differently. It was a raw conversation . . . He was upset. He felt as though that ruined his chances."

Junior and Megan could have these types of conversations because theirs was an "adult" relationship based on respect and transparency, unlike his other postdivorce unions.

"When you're long distance, you're kind of forced to really talk a lot—if it's going to work," she said. "I'm 38 years old, I'm not one of those young people. I wanted a partner that I could talk to and tell my highs and lows of the day to, and he was that person. He would share with me as well."

Junior's mood was much lighter the next day, May 1, when longtime friend Joey Stabb phoned him. The two had been workout partners for eight years, beginning in the mid-1990s. They'd meet at 6:30 each morning at Gold's Gym in San Diego. They fined each other $20 for being a minute late and $50 for failing to show up.

"Joe Joe! What's going on?!" Junior said. "We're having a big party at the restaurant on Saturday. Let's meet at 12 o'clock. I got the fight

[Floyd Mayweather versus Miguel Cotto], and we got the Kentucky Derby going."

"He was Junior," Stabb said.

The rest of the day was spent lounging and hanging out with Megan. Junior also sent text messages to all his kids and to Gina telling them that he loved them. He also made plans to meet his father at the restaurant the following morning. Sadly, he never made it there.

18

"9-1-1 Emergency"

TYPICAL OF MAY mornings along the San Diego coastline, the second day of the month began with marine layer clouds that blotted out the sun and suppressed the temperatures. The morning was lazy and tranquil, with gently crashing waves providing a soothing soundtrack.

Megan awoke around seven o'clock and headed to the gym 45 minutes later. Junior was still in bed in the master bedroom. He didn't join her because he had plans to meet his father.

At about 9:15, Megan left the gym and phoned Junior four or five times. He didn't answer, so she drove by the gym where he typically worked out. His car wasn't in the lot and there was no sign of him through the windows, so she headed back to the house and entered through the garage.

It was eerily quiet inside. She called out to Junior, but he didn't answer. When she got upstairs, she noticed Rock, Junior's 125-pound pit bull–mastiff mix, in the living room. *That's unusual,* she thought to herself. Rock wasn't normally in the living room by himself.

After walking down the hallway, past two spare bedrooms, she didn't see Junior in the master bedroom and turned around to retrace her steps. That was when she noticed that one of the bedroom doors was shut. That was unusual as well.

The room belonged to Sydney, and when she entered she saw Junior slumped on the queen-sized bed. His body was limp, and there was a gun next to it. She immediately picked up the phone and called the police.

9-1-1 emergency!

"My boyfriend shot himself! Oh my God! Oh my God! My boy-friend shot himself!" a crying and gasping Megan screamed.

"Do you know if he's breathing?" the operator asked.

"I don't think so . . . What do I do?! Oh my God! How do I tell if he's breathing, ma'am?"

"Can you check his pulse? See if his chest is rising?"

"It's not rising."

"Where did he shoot himself?"

"I can't tell, ma'am. It looks like in the heart."

Megan was frantic. But her mind quickly turned from Junior to his kids.

Oh my gosh, she thought to herself. *The kids. I've got to get to the kids and Gina.*

She had never met Gina, but she respected her because "Junior respected her."

Within an hour, TMZ reported that Junior was dead from an apparent suicide. It was the third time in 15 months, and the second in two weeks, that a retired NFL player had killed himself with a bullet to the chest. But the suicides of Duerson and Easterling did not carry the same impact as that of Junior. They were good players; he was an icon. And his death buckled the knees of the National Football League, from San Diego to New England.

"We all lost a friend today," Chargers president Dean Spanos said in a statement. "This is just such a tragic loss. One of the worst things I could ever imagine."

"He may have been one of the most charismatic Patriots players in franchise history," said team owner Robert Kraft. "Today, the fans of the teams for which Junior played — San Diego, Miami, and New England — lost more than a legendary football player. We lost our 'Buddy.'"

As the news spread that morning, Papa Seau had no idea of what had transpired. He had gone to the restaurant, but was driving home after his son failed to show. Along the way one of his daughters called and told him to come by Junior's house. No reason was given — just get there as fast as possible.

Mama Seau was in church when she received a similar call. Again, no reason was given.

When she arrived at his home, a crowd of cars, people, and TV crews lined The Strand outside and above Junior's house. She knew something wasn't right, but what?

The news that he had taken his life didn't fully register with her when her daughters told her. It was almost like that day in the Chargers' meeting room when Jim Vechiarella, the linebackers coach, warned everyone that things weren't going to end well for Junior. You hear it, yet you still have a hard time processing it.

It was not until Mama Seau stepped before reporters and disbelieving fans outside the house that the weight of the moment hit her. She was composed initially, but eventually her emotions took control.

"I don't understand, I don't know anything . . . ," she said. "I'm shocked. But I appreciate everybody. Show your love to my son."

Then came the tsunami of grief.

"Where's Junior?" she said between gasps and tears. "Junior never do nothing to you guys. But I say today: Thank you. I appreciate you guys show[ing] your love to my son."

Finally, she broke down completely.

"I don't understand who do this to my son, but I pray to God, 'Please, take me! Take me! Leave my son!'" she said. "But it's too late. Too late . . . He never say nothing to me. Junior! Why you never tell me you're going?! I pray to God: 'Take me! Take me! Leave my son alone.'"

Mama Seau then collapsed in the arms of her children. Later, inside the home, Junior's family and friends gathered for a Samoan ceremony over the body. Amid the tears and shock, they recited Psalm 23: *The Lord is my shepherd; I shall not want . . .*

Each then had a private moment with the man who had meant so much to everyone. Some reached down and touched him. Bette Hoffman, the longtime director of his foundation, and a woman whom Junior affectionately called Mom, leaned over and kissed him.

Finally, the coroners placed the body on a stretcher and rolled it out to a waiting white van, where it was loaded into the back. The van drove along the shore before turning inland, toward the highway. There was a hope that he finally was at peace.

Fans and community members continued to drive, bike, and skateboard to The Strand, as if needing confirmation that Junior had passed. They also made pilgrimages to his restaurant.

That afternoon Paul Sellers was preparing to head to Petco Park for a Padres game. He had on a Padres jersey and cap and was prepared for afternoon baseball. But instead of spending the day downtown, he found himself sitting outside Seau's The Restaurant, rosary beads in his hand.

"I saw the news on TV and said, 'I've got to go down there,'" he said. "I wanted to say a prayer for him and his family."

He was joined throughout the day by a steady procession of fans who felt like they had lost a friend. They left flowers, cards, candles, teddy bears, and pictures of Junior outside the restaurant. The scene was much the same at Junior's home. In the days following his death, people left Chargers "55" jerseys with notes written in black ink on the white numerals. Others tied helium balloons in the shape of his jersey number to the metal barricade that was set up to keep people from getting too close to the house.

On one barrier outside his garage, someone left a paper coffee cup. Written on the side: 2 PUMP VAN[ILLA] LATTE. A BET'S A BET. Presumably that was Junior's regular order.

In the sand across the street from Junior's home, Jimmy Garcia used small rocks to create a large cross with a "55" and "RIP" beneath it. The public opened its collective heart for Junior because he always gave his heart to them.

"A little over a year ago, a good friend of mine was stabbed to death outside of a bar in Carlsbad," Joshua Donahue wrote to the local paper. "The next weekend we held a memorial fundraiser with a raffle to raise money for his 4-year-old daughter. Junior showed up to the memorial with an autographed football he wanted to donate for the raffle. This is just one of the countless examples of the kind of person he was, constantly thinking of others, always reaching out to the community. I don't think there is a single soul in Oceanside he hasn't touched in some way."

"Like many of you I was a fan of Junior Seau the football player," Glenn Encarnacion wrote. "But it was a gesture of kindness that really made me a fan of the man. When a friend of mine, a huge Chargers fan, had a terrible accident with severe injuries, I wrote Junior and asked him if he could simply sign a card. He did that and so much more, visiting my friend and showering his family with Chargers and 'Say-OW!' related gifts. Thank you, Jr. RIP."

"I'm shocked and heartbroken," said John Carney, a former teammate. "Being a friend and teammate of Junior's was a highlight of my career. The positive influence he spread among teammates, coaches, fans, and even opposing teams is unmatched. He'll be greatly missed."

"I'm sorry to say, Superman is dead," Chargers chaplain Shawn Mitchell said. "All of us can appear to be super, but all of us need to reach out and find support when we're hurting."

Junior's death sent a current of depression through the county. Questions outpaced answers by a significant margin. Everyone wanted to know how someone so outwardly happy and positive, so universally beloved and respected, could take his own life. Some even refused to accept that he actually did it, going so far as to speculate that someone else pulled the trigger.

Conspiracy theorists wondered why the memory card in his phone was missing. And where did Junior get the .357-caliber Magnum revolver? He wasn't known to like or own firearms.

Others wondered if drugs might have been a factor, although autopsy and toxicology reports showed no alcohol or common drugs of abuse in his system. They did detect 0.14 milligrams of zolpidem, the drug Junior took to treat his insomnia, and a trace amount of naproxen, a nonsteroidal anti-inflammatory, that was "consistent with therapeutic use," according to deputy medical examiner Craig Nelson. But there was nothing else out of the ordinary.

However, a study of his brain did show that he was suffering from CTE, the degenerative brain disease that's brought on by repeated head trauma. Duerson and Easterling also were found to have suffered from the disease, whose symptoms include mood swings, impaired judgment, and struggles with impulse control. The family was so alarmed by the findings — and mounting allegations that the NFL hid the dangers of concussions from players for decades—that they sued the league in hopes of forcing it to fully disclose what it knew and when it knew it regarding the disease.

Why did Junior Seau kill himself?

"No one is going to know but Junior," said Leonard Mata, an Oceanside native and a member of the city's police department.

The truth is that the answer is much more complicated than A + B = C. In addition to suffering from alcoholism, depression, prescription-drug dependency, financial issues, family strains, and brain trauma—a Molotov cocktail of ingredients by themselves—Junior also suffered from a broken heart. He often seemed troubled that he couldn't consistently be the man everyone perceived him to be, or that he wanted to be.

On his kitchen counter following his death, authorities found a piece of paper on which he had written the lyrics from one of his favorite country songs, "Who I Ain't," cowritten by his friend Jamie Paulin, a Nashville-based songwriter. The lyrics spoke to the internal struggle of a good man who regrets having done bad things.

As much as some viewed Junior as a myth, he was merely a man. He bled when cut, cried when sad, and succumbed to temptation even though he knew it was wrong. He would give all that he had to teammates and friends, yet neither ask for nor accept anything in return. He was a humble kid with a special gift, simultaneously pushed by his family and driven by himself. His shoulders were broad because they needed to be. He carried a lot of responsibilities on them.

"I often think about how Junior was the leader and the captain of the team, the face of the franchise, and how there was so much pressure on him," said LaDainian Tomlinson, who likened Junior's passing to the loss of an older brother who teaches you how to be a man. "People can say there isn't a lot of pressure that goes along with that, but there is. You're the one that not only the public is looking at to be perfect, but also the people in the organization and your teammates. Junior was always that guy.

"Nobody does this if everything is just fine or things are going great. I feel awful that Junior didn't feel he was close enough to anybody that he could say, 'Look, something isn't right.' We all need someone we can go to and say, 'There's something going on with me.' But that's who Junior was—he didn't want us to know he was hurting on the field, so off the field he certainly wasn't going to say anything."

The struggle to transition out of football is real for countless players. Following Junior's death, several of his teammates admitted that they had also considered suicide, because their lives suddenly seemed to lack purpose or direction. Orlando Ruff, the undrafted linebacker whom Junior took under his wing, was one of them.

"It's a big adjustment," he said. "It's one of those things that since you were a little kid all you've known is football. Sure, you've been exposed to other things, but your focus has been narrowed on football, and your focus is what gets you to the highest level. Everyone always talks about life after football and how you should plan—and that's very true. But when you're in the heart of it, you're not thinking about what's after it. You're thinking about being the best at what you can do right now. Part of what makes you who you are is your feeling of invincibility, that it's never going to end. But the reality is that not many of us get to write our own ending and how we're going to go out. When it happens, you're like, 'Okay, I've got to move forward. I'm okay with this. I can deal with this.' But oftentimes we're not ready.

"We're good with the X's and O's over here in football," Ruff continued, "but now we're in unfamiliar territory. We're in the real world. We don't necessarily have all the comforts around us that we once had to get us through whatever dilemma we might be facing. We'll follow our passion, but we may not have all the details and information and training necessary to succeed. We say it's okay and that we'll learn on the go, but then we go out and we fail, something that has never happened to us before. Then we start to question ourselves . . . You're kind of left out in the cold, and you don't have a support staff, someone you can call. You're used to being up here and now you're down here. It's a tough adjustment. A lot of us don't make it out of that."

Ruff keeps private the details of his situation, but in hindsight he could see himself slipping gradually into an abyss of uncertainty. "I'll put it this way," he said. "I saw some dark places in the three years it took to find my direction."

His guiding light flashed while fighting the NFL on a worker's compensation claim. The attorney representing him was going over the case, and after a while Ruff asked him how he became a lawyer. Suddenly, he discovered his purpose for carrying on.

"I said, 'Wait a second, this sounds like something I can do,'" Ruff recalled. "My case was resolved, and later that day I went and got an LSAT book. I didn't know what I was getting into, but figured I had tried everything else and it hadn't worked, so let me give this a try. Signed up for the LSAT, took it, went to law school, and graduated. I didn't know what I was getting into, but then I realized there really was a purpose for me to help others. That's when I said, 'I'm supposed to be here.'"

During the two-and-a-half-year span from February 2011 until September 2013, at least five active or former NFL players committed suicide: in addition to Duerson, Easterling, and Junior, there were Kansas City Chiefs linebacker Jovan Belcher, who killed his girlfriend before taking his own life, and former Chargers defensive back Paul Oliver. Only the individuals know the reasons for their actions, but the national attention led the NFL to launch a crisis support line in July 2012. It was established to help active and retired players and their families and is supposed to operate independently from the NFL, with all calls being confidential.

"There is no higher priority for the National Football League than the health and wellness of our players," Commissioner Goodell wrote in a letter to team personnel and fans, even as the NFL was being sued by thousands of players for allegedly hiding the potential long-term dangers of concussions.

Before Junior's death, Aaron Taylor had not heard from his former teammate for more than a year — not since they last attended a 12-step program together. On the morning of May 2, 2012, he was at the Pacific Athletic Club in San Diego when his cell phone rang. A mutual friend was calling to ask if he had heard the media reports that Junior might have killed himself. Taylor could only shake his head. He was sad but not shocked, because he'd had a sense that the worst could be coming — like an offensive lineman who whiffs on a block, then turns and sees the defender closing on his quarterback. He knows that, with few exceptions, the outcome is a fait accompli. The only question is, how long before the sack occurs?

"I do a lot of work with retired players now, trying to help them transition," Taylor said. "I know what the cycle is and all the dynamics that contribute to it. It's a perfect storm of enabling that ends up slicing our throats. When we get in the real world, we defer and revert to all the things that helped us as professional athletes, and they just don't work: *Don't think — react. Never admit weakness or ask for help. Push through, suck it up, fight on.* It's all those things that allow us to do what we do. But looking back at Junior, he had enough strength and toughness to be able to deal with and to play with broken bones, but not enough to deal with his emotions and life on life's terms."

Even today some of his former teammates and friends remain angry that more was not done to help Junior. Rodney Harrison is among those who scoff at the notion that it was impossible to see the life-threatening road on which Junior was traveling.

"I was surprised that he would do something like that because I knew how much he loved life and how deeply he cared for his kids; he always talked about his kids," Harrison said. "But it didn't surprise me in that I knew his pain ran deep. Junior couldn't be here and have people laugh at him or look at him for not being able to take care of his family. So it didn't surprise me in that sense. Based on the last few years of his partying ways and the [emotional] distance and just him not returning calls — he wouldn't call me back, or the conversations were always short. I knew things weren't right just from the stories that I heard from a distance that he was on a destructive path. He was hurting, he was hurting real bad.

"Junior carried such a facade, but there was so much pressure on him that he always had to be Mr. Perfect, that he always had to take care of people, that he always had to present himself as being something. He could never relax and be at peace. At one point it seemed like he was at peace, and I think that divorce really, really hurt him. The two things that he told me that really gave him peace were his children and that surf, to be able to go out there and surf those waves. They gave him that peace for that moment. But outside of that, he was struggling with a lot of demons."

Demons that those in his San Diego–based inner circle should have been able to see, in Harrison's eyes.

"There were a lot of people around Junior that knew Junior was going through certain things, but they didn't care," he said. "They cared about Junior because he gave them something. He gave them a sense of importance, a good time: *I'm hanging out with Junior Seau.* The people close to him who say they didn't see the signs — you've got to be joking, you've got to be kidding me. I didn't even hang out with Junior and I could see the signs, I could see the depression. I tried to reach out to Junior, and that's the problem; he surrounded himself with people that . . ."

Harrison didn't finish the sentence. Then again, he didn't need to. Is it a coincidence that some of Junior's closest party buddies declined to be interviewed for this book, most notably Jim Barone and Ken Ramirez? They said it stemmed from a pact they made to keep their memories private and protect his memory. But could it also have been because they felt some guilt about not having challenged him on the behaviors they were witnessing?

For instance, instead of demanding that he seek help after driving off the bluff, they followed Junior's lead and made light of the situation. They even took to calling him "Cliff" instead of "Bug" or "June." Another time, while playing golf together, Junior suddenly turned angry and threatening toward Barone, but was back to being fun-loving Junior in the relative blink of an eye. The change in personality frightened Barone, who told him: "You need to get your head checked out." It was said in such a way that everyone laughed, but the reality is that Junior did need to be checked out. But none of his party friends pushed him to do so, perhaps because they knew Junior might view that as negative energy and distance himself from them.

"If there were negative people around him, he would be through with them," Harrison said. "He would not have them around him. For people to sit there and say they didn't see the signs, it's a crock. He had people that enabled him. He had people that laughed at his jokes, people that partied with him and wanted a good time with Junior Seau. But the people close to him knew he was hurting. No one drinks like that. No one beats themselves down like that, into oblivion. No one gets into fights. That's not the Junior that we know. But they turned their head to it because he gave them a good time. He made you feel special. He was endearing, engaging. He made you feel important. You could be the guy working at the post office and he'd make you feel like you were making a million dollars. That's the special quality that he had."

19

Gone but Not Forgotten

THERE'S NOTHING THAT distinguishes the house from others on the block. It's a two-story stucco with a three-car driveway and a brown welcome mat at the front door.

It's not until you step through the double doors that you realize the home belongs to Junior's parents. On one wall is a large poster-like painting of Junior in his Chargers uniform, crouching low, running forward, mouth open, preparing for contact. The caption reads: "When Lightning Strikes."

Upstairs, on a shelf above the double doors that lead to the parents' bedroom, sit two game balls and a helmet with a Super Bowl logo. On a table downstairs, there's a framed picture of Junior and his father entering Qualcomm Stadium the night the Chargers beat the Steelers to earn a trip to their only Super Bowl. Papa Seau is wearing a Chargers pullover, a Hawaiian lei, and a pride-filled smile in the picture. Junior is wearing a dark pullover and an AFC Champion hat. He isn't smiling, but there is a look of contentment on his face.

"I miss him," Mama Seau said from a love seat in the living room, nearly two years after Junior's death.

She has learned to deal with his absence, but the pain is never far away. In the months after his passing, she would travel roughly two miles daily to his gravesite and spend hours sitting and talking to him. It was her way of coping. For others, including his kids, counseling sessions helped them get through the pain.

"Honestly, it was confusing and hard growing up as his only daughter," Sydney said. "There was just so much pressure. I wasn't a boy, but

I felt just as much pressure as my brothers did. I wanted to be better. Because he was gone all the time, I just felt that there had to be something I could do to make him proud enough to just stay a little while, to be with us kids. It was a competition within myself—not with my brothers or my family name—just with me to prove that I was good enough to be his one and only daughter. It was overwhelming because he was the only person I wanted to make happy, because I knew he needed it more than anyone else. No one really gets that."

Sydney paused and gathered her emotions.

"He was the light in my life," she continued. "There's not a day I don't think about him, I don't miss him. But there has to be some way that I can reconcile with what happened, and it's that I know I did everything I could to keep him here. There were things that were out of his control and my control [i.e., CTE] to keep him here. But I really do think he made it. He did everything he could in his 43 years, and he made a difference and an impact on hundreds and thousands of people. A lot of people can't say that."

Prior to this book, Junior's parents and other members of the Seau family refused to speak publicly about him following his death. They were hurt and disappointed by stories that focused on only the worst parts of his life, and they feared that his children might come to think less highly of him.

"I hear a lot of negative stuff about how Junior treats women," his sister Mary said. "The way he grew up with my sister and I and my mom—Junior loved my mom, he loved me. He always treated the women differently, with respect. He's always been happy, outgoing, willing to do things to help others in the family. He was so family-oriented. He's been down, but he's very sensitive in some ways. If I could tell his kids anything, it's that your dad was an awesome dad, an awesome brother. He always wanted to do right. Whatever you hear negative out there . . . that's not him. He believed in God. If you'd have seen him growing up at church—we'd have a luau and he'd put his whole energy into dancing. He and Sal [Aunese] would dress up as women in bras, etc., and dance and have fun. He was just a loving person."

Although history will show that Junior Seau took his life, his legacy will reflect a man who gave more than he took. It was important to him to make people happy. It was why he allowed Eric Olsen, then a nervous and unsure high school kid, to block him to the ground at a football camp. It was why he went to city hall and took the certification test when former Chargers team orthopedist David Chao and his fiancée asked him to preside over their wedding. It was why he showed up the day after tearing a biceps and spoke to the elementary school class of Allison Yahnke, the daughter of his financial planner, Dale Yahnke. Not only that, he began his talk by telling the kids about a special buddy he had made 30 minutes earlier. Her name: Allison Yahnke.

That was Junior's way. He was not the type to simply give money to charities; he believed in connecting with people. Donating his time was more fulfilling than donating his money. He even created an off-season beach boot camp for Oceanside kids who could not afford personal trainers — and he often trained them himself. When other community members started showing up — young and old, male and female, gang-bangers and non-gang-bangers — he opened the sessions and his arms to them as well.

Years after his death, his ability to touch people can still be seen at his gravesite, where parents leave pictures of their kids with Junior. In one, a young boy is wearing a helmet and being held by the star linebacker. It is debatable whose smile is wider.

There's also a handwritten note wrapped in protective clear plastic. It reads: "You mean more than you know and are missed but not forgotten." It's a thought shared by an entire community.

Junior ended most public speeches by reminding the audience to "live for today, build for tomorrow, pray for the rest." He always sought to find light where there was darkness, a quality he passed on to at least one of his children, as reflected in a school essay written by his youngest child, Hunter, in the fall of 2014. It was titled "Life Is Valuable" and read as follows:

> May 2nd, 2012, was the worst day of my life. This was the day that my Dad unexpectedly passed away. It's been just over two years, but the scars will forever remain.
>
> Sadly, I may have only seen him once a month for a weekend here and there prior to that day, which was never really enough. He spent

the majority of his time playing professional football and was rarely in town.

I'll never forget the weekend just weeks before he died. It was in April of 2012 and I was at my Dad's house and we were getting along perfectly. We did all of our favorite things together: We went golfing at Del Mar Country Club, then we went to my favorite restaurant, SEAU'S. The next day we went in the ocean and hung out together. It was the best weekend I had ever spent with my Dad. It was just he and I.

Then on May 2nd I heard about my Dad passing away. My mom rushed to get me from my school that morning with my older brother and sister in the car. Everyone was so sad and crying. I cannot put it into words how painful it was to hear my Dad was not here anymore. I really didn't know how to react. I was so sad, but I wasn't crying. It was like there was a knife that went through my heart and it just hurt. The pain just would not go away, and it was a pain I had never felt before—and by far the worst.

Everything from May 2nd to the end of May was just a blur. I can't remember much of anything that happened in between. It was like I was in a daze of sorrow and I couldn't get out of it. My Dad and I talked about going to Hawaii and all these elaborate things that we wanted to do together. That is why I didn't understand why he would leave me so soon. Sadly, I don't think I will ever understand.

My Dad had the best sense of humor and was always positive and kind. He would always tell me, "Son, remember to always stay humble." He never wanted to draw attention to himself. He was the most humble person I have ever met. After football games he would go out to the crowds and sign things for hours, even though he had things he would much rather be doing. But he would sign things for the sake of his fans. It shows what a kind and considerate man he was.

He also told me, "Son, you need to always be grateful for what you have." My Dad grew up in a very poor family and then grew up to be a very successful man, but he worked hard to get what he had. He was the person I looked up to and I've always wanted to be somewhat like him.

Eleven years prior to my Dad's death, my Mom and Dad got divorced. This was also a sad time for my family. Even though I was only 1 years old, this has affected my life dramatically [because] I had to switch off parents. I was almost always with my Mom, but on occasion I would go to my Dad's house.

Both of these tragic things happened so quickly, and they both had a positive and negative effect on my life. The negative effect is my Mom and Dad split up, causing me to grow up with one parent. And later my Dad passed away, which has made my life harder because I no longer have my Dad around at all.

If there is anything positive, it is that I've learned how valuable life really is, how we need to love and cherish those around us. People don't often realize how things can happen so quickly and unexpectedly. Now I value everything I do and try to pick out the good in everything because you never know when something bad can happen. Why not enjoy the good in life while you still have the privilege to do so?

Epilogue

On January 31, 2015, the Pro Football Hall of Fame selection committee gathered in a ballroom on the second floor of the Phoenix Convention Center. The 46 members were there to determine the Class of 2015, which would be no simple accomplishment considering they had to reduce a who's who list of 15 modern-era finalists to a maximum of five. (Two contributors and a seniors nominee also were up for consideration, but the vote on them would be conducted separately from the modern-era candidates.)

One by one the finalists were presented for discussion, starting with the running backs (Jerome Bettis and Terrell Davis), then the linebackers (Kevin Greene and Junior Seau), then the one quarterback (Kurt Warner), and the coaches (Don Coryell, Tony Dungy, and Jimmy Johnson). They were followed by the sole kicker (Morten Andersen), the wide receivers (Tim Brown and Marvin Harrison), the safety (John Lynch), the defensive end (Charles Haley), and the offensive lineman (Orlando Pace).

The discussion for each could go as long or as short as the committee desired. It turned out that all but one candidate was discussed for 12 minutes or longer. The lone exception was Junior Seau, who held the floor for only seven minutes.

Truthfully, it was surprising the discussion of his candidacy went that long, considering that he is regarded as one of the greatest linebackers in the game's history. No one would have blinked if the debate began and ended with the mention of his name. But it went for seven minutes because some committee members wanted to recount their personal memories of watching him perform. Their words were similar to those of ESPN analyst and former Broncos Pro Bowl linebacker Tom Jackson the next morning—that he couldn't take his eyes off Junior because Junior played every play as if it was his last one.

Junior was the type of player for whom scouts checked all the boxes when evaluating linebackers. Not only was he the first in line when the football gods were handing out size, speed, quickness, and strength, but he also was blessed with the intangibles of greatness: passion, discipline, accountability, and an unrelenting work ethic.

"There was a guy that I wish I had kept some of the film of from our time together," said Wisconsin coach Paul Chryst, who was an assistant with the Chargers from 1999 to 2001. "Just a great player, but never afraid to get knocked on his ass. In college it's a badge of honor to say you've never been knocked on your ass, but that means you've never cut it loose and played on the edge. He did that.

"I remember asking him in off-season workouts, 'Junior, what do you get out of this? Why do you go full speed?' He said, 'I don't get anything out of it. But the linebackers need to learn how to play with me.' He didn't say it in a cocky, egotistical way. I thought he was a pro's pro. He was something special."

When the Hall of Fame vote took place, Junior's family was spread across the world. Sons Tyler and Hunter were in Phoenix with Gina and Bette Hoffman, the executor of his estate. Daughter Sydney was in London, as an exchange student from the University of Southern California. Son Jake, a freshman at Duke, was at school preparing for his first lacrosse game of the season. And his parents were home in Oceanside, awaiting word of the fateful vote.

When his spot in the Class of 2015 was official, emotions were mixed. "The happiest sad day you could imagine," said Hoffman.

Junior's dream was always to have a bronze bust in Canton, Ohio, but the reality that he would not be there when the cover was lifted off hit everyone with the force of a blindside tackle.

"I'm so disappointed that I'm going to be giving a speech to an empty chair in Canton, but I understand why," said Sydney, who will present Junior to the audience at the induction ceremonies in August 2015. "He should be here, but he's not for a reason, and that's what we have to take from it and keep going and growing. All I'm focusing on is being honored to introduce the light of my life. Yeah, I'd love for him to be here and make fun of my amateur speech writing and tell me to stick to my day job, because I know that's what he would've done. And yeah, I'd like to hear him say, 'All right, Syd. Dreams are for free. You didn't have to do that. You can take a seat. Have a great day.' I know he would say that because I can hear him saying it right now."

Sydney paused and laughed. She could feel her father's presence, even if he wasn't there.

Acknowledgments

I am incredibly indebted to Junior's family and friends, some of whom broke the silence they had maintained since his death to help me paint a fuller picture of his life and legacy. Their confidence that I would present their memories with honesty and fairness impacted every word and page in this book.

It would be disingenuous to say the project wasn't more challenging than I anticipated, largely because of the sensitivity of the subject matter and the friction between Junior's parents and siblings and his ex-wife Gina. From the outset each side wondered whom I would be working with in telling his story. What they really were asking me was, "Whose version of the truth are you going to tell?"

With Junior, there were multiple versions. He hated drama and therefore would tell people what they wanted to hear because keeping the peace was paramount. However, by telling conflicting versions of the same story, Junior only fanned the tension-tinged flames and created a divide that increased dramatically following his death.

Thankfully, his family and friends agreed to trust me with their memories of him because they knew Junior and I had had a close professional relationship that later evolved into a friendship. We never hung out together, but my two daughters and I were regulars at his foundation's functions. I always marveled that the man's heart was as big as his persona.

That relationship was the primary reason why his family, after two years of grieving, finally accepted my phone calls and answered my knocks at the front door. His sisters, Mary and Annette, were instrumental in filling the voids in not only his early years but also his last months. His parents, through broken English, provided perspective and clarity. And his trainers with the Chargers—Keoki Kamau and James Collins, who were as close to him as his siblings—pulled back the curtains to reveal how a man could seemingly will himself through injuries that should have sidelined him.

Gina Seau played a key role in this book by providing access to the private journals Junior kept during his career—he often wrote on paper what he refused to say to the media or even his friends—and then by recounting their days together after he drove his SUV off a bluff.

Daughter Sydney and son Jake were incredibly honest and raw when discussing their relationship with their father, who loved them tremendously but struggled to show it. And son Hunter showed incredible generosity by sharing the class essay he wrote about the day his dad died. His words were simultaneously heartbreaking and uplifting.

Megan Noderer, the girlfriend who found his lifeless body the morning of May 2, 2012, showed great courage in speaking with me. For three years she had declined to speak with anyone but Gina Seau about that morning. But in February 2015 she opened up to me about her relationship with Junior and the first thoughts that went through her mind when she couldn't resuscitate his body on that horrific morning.

I also owe thanks to the hundreds of former teammates and friends who spoke to me about Junior before and after his death. I began covering him as a reporter for the *San Diego Union-Tribune* in 1996 and wrote countless stories about him. That's why this book didn't require a tremendous amount of "research," because I basically have had a sideline view of his adult life.

One person who earned my utmost respect was Rodney Harrison, Junior's teammate in San Diego and New England, because he said publicly what few others had the courage to voice—that close friends of Junior's who claim they didn't see he was on a destructive path are kidding themselves. They were not friends, they were enablers.

I'd also like to thank family and friends for their support and wisdom, notably, Dan Wetzel, Farrell Evans, Bruce Feldman, and Lars Anderson. And lastly, I have to give a shout-out to Houghton Mifflin and editor Susan Canavan for having the conviction to not only pursue this project with a rookie book writer but also treat it with the sensitivity it deserves.

Notes

INTRODUCTION

Author's Interviews

Billy Devaney, Mike Riley, Nick Saban, Bill Belichick, Rodney Harrison, Mark Davis, and Drew Brees.

1. "I HAVE TO BE BETTER THAN ME"

Author's Interviews

Sai Niu, Tiaina Seau, Luisa Seau, Annette Seau, Mary Seau, Pulu Poumele, Melissa Waldrop, and Don Montamble.

Other Sources

Daniel de Vise, "Seau Enjoying Retirement — Senior Seau, That Is," *San Diego Union-Tribune,* December 14, 1996.

Don Norcross, "Junior Seau: He Channeled His Energies," *Evening Tribune,* June 10, 1987.

Tom Shanahan, "Seau's Best Average Is in Classroom," *Evening Tribune,* February 11, 1986.

2. SUCCESS AND SHAME: ONE AND THE SAME

Author's Interviews

Don Montamble, Sai Niu, Tiaina Seau, Luisa Seau, Annette Seau, Mary Seau, Clarence Shelmon, Melissa Waldrop, Randall Godinet, Gary Bernardi, Mark Carrier, Craig Hartsuyker, and Pulu Poumele.

3. "THERE WAS NOTHING JUNIOR ABOUT HIS GAME"

Author's Interviews

Leroy Holt, Tim Ryan, Michael Williams, Mark Carrier, Craig Hartsuyker, Bobby April, Tom Roggeman, Melissa Waldrop, Annette Seau, and Mary Seau.

Other Sources

Chris Dufresne, "Junior Seau's Brief, Ferocious Career at USC Is Recalled," *Los Angeles Times,* May 3, 2012.

4. FERDINAND THE LINEBACKER

Author's Interviews

Steve Feldman, Gina Seau, Melissa Waldrop, Annette Seau, and Mary Seau.

5. DREAM TURNED NIGHTMARE

Author's Interviews

Sid Brooks, Burt Grossman, Billy Devaney, Bobby Beathard, Steve Feldman, Dave Pearson, John Clayton, Dan Henning, Tiaina Seau, Mary Seau, and Gina Seau.

6. "I'VE GOT TO GET BETTER"

Author's Interviews

Steve Wisniewski, Gill Byrd, Leslie O'Neal, Vencie Glenn, Keoki Kamau, Dan Henning, Steve Feldman, Gina Seau, Burt Grossman, Mary Seau, and Martin Mayhew.

Other Sources

Clark Judge, "Seau Takes More Shots at McCants," *Evening Tribune,* October 25, 1990.

Letters to the editor, *San Diego Union,* September 9, 1990, and September 16, 1990.

T. J. Simers, "He Fights On: Junior Seau Uses Memories of Being Treated as Outsider at USC as Inspiration in NFL," *Los Angeles Times,* September 1, 1991.

7. THE TURNAROUND

Author's Interviews

Dan Henning, Chris Fore, Gina Seau, Burt Grossman, John Fox, Billy Devaney, and Orlando Ruff.

Other Sources

Tom Krasovic, "Seau Is Still a Fan, Keeps Returning to Football Roots," *San Diego Union,* November 2, 1991.

Junior Seau journals.

T. J. Simers, "He Fights On: Junior Seau Uses Memories of Being Treated as Outsider at USC as Inspiration in NFL," *Los Angeles Times,* September 1, 1991.

T. J. Simers, "Analysis: A Hard Start to Ross' Reign," *Los Angeles Times,* October 2, 1992.

8. SUPER FOLD

Author's Interviews

Gina Seau, Keoki Kamau, Tiaina Seau, and Luisa Seau.

Other Sources

Kevin Kernan, "Local Boy Seau Makes Good on His NFL Goal," *San Diego Union-Tribune,* January 16, 1995.

Kevin Kernan, "To Win, Seau Will Have to Work 2 Jobs," *San Diego Union-Tribune,* January 28, 1995.

9. CHEATING . . . IN BUSINESS AND MATRIMONY

Author's Interviews

Junior Seau, Billy Devaney, Gina Seau, Dale Yahnke, Keoki Kamau, Annette Seau, and Mary Seau.

Other Sources

Junior Seau journals.

Kevin Kernan, "To Win, Seau Will Have to Work 2 Jobs," *San Diego Union-Tribune,* January 28, 1995.

Jim Trotter, "Seau Has to Adjust to Forced Inactivity," *San Diego Union-Tribune,* August 7, 1997.

Jim Trotter, "Not Pretty, but a Win," *San Diego Union-Tribune,* September 8, 1997.

Jim Trotter, "Bottled Inside, Seau Frustrated by Lack of Rush," *San Diego Union-Tribune,* September 25, 1997.

10. THE ANTI-LEAF

Author's Interviews

Junior Seau, Bobby Beathard, Billy Devaney, Rodney Harrison, Mike Riley, Paul Chryst, Terrell Fletcher, June Jones, Gina Seau, Bette Hoffman, Pisa Tinoisamoa, and Joe Pascale.

Other Sources

Jim Trotter, "A Miffed Seau Is Workout No-Show," *San Diego Union-Tribune,* June 10, 1998.

11. "MY HEAD IS ON FIRE!"

Author's Interviews

Junior Seau, Mike Riley, Joe Pascale, Paul Chryst, Orlando Ruff, Derrick Brooks, Billy Devaney, Peyton Manning, Bill Belichick, Gina Seau, and Fred McCrary.

Other Sources

Junior Seau journals.

Jim Trotter, "Chargers Preseason, Chiefs Burst Hopes for a First Win," *San Diego Union-Tribune,* September 4, 1999.

12. THE TRADE

Author's Interviews

Junior Seau, Marty Schottenheimer, Dean Spanos, Drew Brees, Orlando Ruff, Rodney Harrison, Buddy Nix, Bette Hoffman, Randall Godinet, June Jones, Marvin Demoff, Liba Placek, and Aaron Taylor.

Other Sources

"Speaking of Seau" (compiled by staff writers), *San Diego Union-Tribune,* March 15, 2003.

13. HELLO, SOUTH BEACH

Author's Interviews

Junior Seau, LaDainian Tomlinson, Peyton Manning, Antonio Gates, Nick Saban, and Dan Quinn.

Other Sources

Jay Posner, "Some Not Pleased by LT's Tribute to Seau," *San Diego Union-Tribune,* October 23, 2003.

Ethan Skolnick, "Seau's Presence Sticks Out," *Sun Sentinel,* May 3, 2003.

Jim Trotter, "Scrutiny to Come with Seau Visit," *San Diego Union-Tribune,* October 24, 2003.

14. THE GRADUATION

Author's Interviews

Junior Seau, Marvin Demoff, Bette Hoffman, Jim Steeg, Dean Spanos, Aaron Taylor, Bill Belichick, Mary Seau, Ronnie Lott, Tedy Bruschi, Dean Pees, Liba Placek, John Lynch, and James Velasco.

Other Sources

WEEI Radio, Boston.

Greg Garber, "Seau Hoping 18th Season Yields the Promised Ring," ESPN.com, January 29, 2008.

15. "THIS IS NOT WHO I WANT TO BE"

Author's Interviews

Junior Seau, Dean Pees, Steve Wisniewski, Ronnie Lott, Gina Seau, Bette Hoffman, Luisa Seau, Annette Seau, Rev. Benson Mauga, Aaron Taylor, Terrell Fletcher, Melissa Waldrop, and Dale Yahnke.

Other Sources

Sam Farmer, "Marcellus Wiley Joins Painkiller Lawsuit Against NFL," *Los Angeles Times,* June 5, 2014.

Nate Jackson, "*League of Denial,* on Concussions in the NFL, by Mark Fainaru-Wada and Steve Fainaru," *Washington Post,* November 22, 2013.

Dan Le Betard, "Jason Taylor's Pain Shows NFL's World of Hurt," January 13, 2013.

Marcellus Wiley, ESPN.com and Associated Press, June 5, 2014.

Jill Lieber-Steeg, "Junior Seau: Bitter End Game," *San Diego Union-Tribune,* October 21, 2012.

Nathaniel Penn, "The Violent Life and Sudden Death of Junior Seau," *GQ,* September 2013.

16. HIGH POINT, LOW MOMENT

Author's Interviews

Gina Seau, Sydney Seau, Jake Seau, Bette Hoffman, Chargers staff, Megan Noderer, Melissa Waldrop, and James Velasco.

Other Sources

Jill Lieber-Steeg, "Junior Seau: Song of Sorrow," *San Diego Union-Tribune,* October 14, 2012.

Matt Potter, "Say It Ain't Seau," *San Diego Reader,* March 6, 1997.

17. "I KNEW HE WAS GOING TO HAVE A HARD TIME WITH LIFE AFTER FOOTBALL"

Author's Interviews

June Jones, Gina Seau, Mary Seau, Annette Seau, Rodney Harrison, Megan Noderer, Mike Norris, and Joey Stabb.

Other Sources

Sam Farmer, "The Junior Seau No One Knew," *Los Angeles Times,* May 5, 2012.

David Leon Moore and Erik Brady, "Junior Seau's Final Days Plagued by Sleepless Nights," *USA Today,* June 2, 2012.

18. "9-1-1 EMERGENCY"

Author's Interviews

Megan Noderer, Gina Seau, Bette Hoffman, Annette Seau, Mary Seau, Leonard Mata, LaDainian Tomlinson, Orlando Ruff, Rodney Harrison, Aaron Taylor, and Fred McCrary.

Other Sources

San Diego police report, May 15, 2012.

9-1-1 call from Megan Norderer to Oceanside Police Department, May 2, 2012.

"Junior Seau: Complete Coverage," *San Diego Union-Tribune,* May 3, 2012, available at: http://www.utsandiego.com/news/sports/chargers-nfl/junior-seau/.

19. GONE BUT NOT FORGOTTEN

Author's Interviews

Tiaina Seau, Luisa Seau, Annette Seau, Sydney Seau, Jake Seau, Gina Seau, Bette Hoffman, Hunter Seau, Mary Seau, Liba Placek.

Other Sources

Hunter Seau, "Life Is Valuable," Fall 2014.

EPILOGUE

Author's Interviews

Gina Seau, Sydney Seau, Jake Seau, Bette Hoffman, Tom Jackson, Keyshawn Johnson, and Paul Chryst.

Index